I0625762

# Smart Buildings by Design

## How the Fourth Trade Connects It All with the Purdue Model

by Jacob Jackson

**Disclaimer**

The information provided in this book is intended for educational and informational purposes only.

While every effort has been made to ensure accuracy, the author and publisher make no representations or warranties with respect to the completeness or applicability of the contents and disclaim any liability for damages arising directly or indirectly from the use of the information contained herein.

Readers should consult with qualified professionals before implementing any strategies described.

---

**ISBN:** 979-8-9988130-0-9

**First Edition: May 2025**

Printed in the United States of America.

To my friends, colleagues, and workmates—

Thank you for the questions, the challenges,
the lessons, and the laughs.

Everything in this book was shaped by the
time spent learning beside you, building with
you, and sometimes fixing what we all
thought would work the first time.

I couldn't have written this without you.

# Author's Introduction

You've probably heard this story before: You show up to commission an Operational Technology (OT) automation project, and there's an HVAC system that talks BACnet, a lighting system that only understands DMX, a fire alarm with dry contacts labeled in Sharpie. You have three vendors blaming each other for why it doesn't work. Meanwhile, the GC wants to turn over the building next week and the owner wants it all integrated into "one screen" and their big Grand Opening ceremony is depending on the tech to make an impression. And the customer is calling you asking why your technician just walked in with a laptop, a bucket and a TV tray and locked himself in one of the mechanical rooms. Sound familiar?

I've lived that story—over and over again, from being:

- the tech with the bucket
- the PM who had to deal with the GC
- the company VP who then has to explain all the schedule slip notices by the construction team that no one responded to when our team was pointing out that we would still need our full schedule duration.

I didn't write this book from the comfort of an ivory tower or an engineering office that never visits the field. I wrote it as someone who's spent decades buried in the guts of buildings, unrolling drawings that don't match reality, tracing unlabeled wires, negotiating firewall exceptions with skeptical IT departments, and trying to explain to executives why their "smart building" doesn't feel that smart. I've encountered nearly every integration challenge out there—and if there are more, I'd prefer not to find them.

Humor aside, my background spans over 30 years in building automation, systems integration, energy monitoring, and mission-critical infrastructure. I've worked on government sites, commercial high-rises, campus-wide networks, entertainment, hospitality, and projects with all kinds of integration. I've been the guy installing panels, writing sequences, troubleshooting point lists, and dealing with three-letter acronyms no one agrees on. I handled client negotiations, hiring and firing employees, contracting subs and vendors, and all of those various bits of working and running a controls business. I've also worked in the different low voltage building trades: communications, fire alarms, HVAC with DDC, power systems, security, audiovisual, lighting controls, and front ends for all of these. I have some international experience, having done projects in Europe and Asia, but this book is US centric. I apologize to my international friends, but I just don't have the experience to write to all locales.

As an industry we have talked about these low voltage systems as the fourth trade for some years. Mechanical, Electrical, Plumbing is considered to be the first three or MEP. The concept of a "fourth trade" evolved organically as building systems grew more complex and integrated. The term started to be tossed around in the 2000's and 2010's as the scope and budget of low-voltage systems began rivaling traditional trades in complex projects (e.g., hospitals, airports, schools), prompting GCs and designers to treat them as a distinct discipline. Today you will hear it called MEPT for Technology. (Why not S for Systems you ask? S is Structural in traditional engineering firms.)

The scope is acknowledged, and our low voltage team has a seat at the project team. Problem solved? No. Each is treated independently. The communications team is told to work with IT and Finishes. They know where the PC, access points, and the desk are going to be, but nothing about where any of the OT needs to go. Each OT trade then develops their own pathways and backbone, which can create a myriad of issues. But the plan is to build a complete tech stack that functions just like the building itself. Access control as the technology doors, BAS as the digital HVAC, Lighting Controls as the digital lights... you get the idea. Bringing in specialist for each trade and their needed technology.

We are going to flip that model in this book. Why? Because the old one's broken. For too long, we've started at the bottom—letting each trade bring its own controls, hoping they'll all magically integrate at the end. That's how you get scope gaps, finger-pointing, and buildings that age badly. We're flipping the script and starting at the top: at the enterprise level, with the owner's goals, the IT department's expectations, and the building's long-term future.

Let's be clear on that. Buildings should last. A commercial building should have a life span of 50 to 100 years, a high rise a little longer at 75-120 years. Even residential buildings should last 30-70 years. The technology in the building? Not even close to that. Audiovisual technology advances so rapidly that a 5- to 7-year lifespan is considered excellent. I've also working in the different low voltage building trades: BAS for HVAC and Electrical we should get 10 to 15 years life span out of.

The point of this book is a top-down integration approach, beginning with enterprise-level strategy and working down through system implementation. The structure emphasizes early-stage integration planning and selling to executive or GC-level stakeholders, followed by implementation details for practitioners. I am going to weave in future proofing, maintenance, and contact with the building operators and owners to keep the building operational over it's lifespan.

This book is a call to the entire low voltage industry—not just HVAC guys, but lighting, security, AV, fire alarm, shading, metering, and IT—to **work together**. If we're going to deliver real integration, we can't keep acting like we're on separate islands. We need common language, shared expectations, and mutual respect.

My voice throughout this book is intentional. It's practical, it's grounded, and yes—it's occasionally irreverent. Because if we don't laugh at this stuff now and then, we'll lose our minds. But behind the humor is a serious message: we can do better. We can build smarter, more flexible, future-ready buildings—if we start working like one team.

So let's get started.

# Contents

# THE PURDUE MODEL FOR SMART BUILDING INTEGRATION

LEVEL 5 — ENTERPRISE / CLOUD

LEVEL 4 — BUSINESS SYSTEMS

LEVEL 3 — SUPERVISORY CONTROL

LEVEL 2 — AREA / FIELD CONTROLLERS

LEVEL 1 — DEVICE I/O

LEVEL 0 — PHYSICAL PROCESS

# PART I: Strategy, Enterprise Integration, and System Architecture

# Chapter 1: Why Building Controls Are Broken—and How to Fix Them

In too many buildings today, control systems are installed in silos—each trade delivering its own narrowly scoped subsystem, with little thought for how it fits into the larger whole. HVAC comes with its own controllers, lighting is managed separately, and fire alarm, security, AV, and shading often have no defined relationship to each other. The result? Overlapping wiring, redundant sensors, vendor lock-in, and missed opportunities for optimization, comfort, safety, and energy savings. But imagine walking into a building where occupancy drives HVAC, lighting, and shades automatically—without requiring eight different interactions. Where dashboards speak a common language, and replacing a controller doesn't mean rewriting the entire BAS.

This book challenges the current model and lays out how to get to a better one.

We propose a new starting point: not with hardware, not with shop drawings, but with integration strategy. Using the Purdue Model as a framework—starting from the enterprise and moving down through facility, controller, and device layers—we aim to help you rethink how control systems should be planned, delivered, and maintained.

### What Is the Purdue Model—and Why Should We Care?

The Purdue Enterprise Reference Architecture (PERA)—commonly known as the Purdue Model—originated in industrial automation, but it has become a go-to framework for organizing control systems in complex environments, including modern buildings.

It structures systems into hierarchical levels, each with its own responsibilities:

**Level 5 – Enterprise / Cloud**
Corporate IT, remote analytics, portfolio dashboards

**Level 4 – Site Business Systems**
Energy management software, CMMS, facility IT servers

**Level 3 – Supervisory Control**
BAS front-ends, integration hubs, central user interfaces

**Level 2 – Area Control / Field Controllers**
DDCs, lighting control panels, security gateways

**Level 1 – Device I/O**
Sensors, relays, actuators

**Level 0 – Physical Process**
HVAC equipment, lighting fixtures, access hardware

---

## Why It Matters

### Common Language
Vendors and trades all speak differently. The Purdue Model gives everyone a shared vocabulary.

### IT Alignment
The Purdue Model mirrors how IT thinks—making security, networking, and data sharing easier to coordinate.

### Smarter Design
Clearly defined levels help avoid redundant systems, tangled integrations, and orphaned devices.

### Future-Proofing
Systems can evolve independently while staying connected and structured.

---

### Bottom Line
If you want a building to think clearly, it needs a clear brain. The Purdue Model is that brain map.

**Author's Note:** The Purdue Model has been widely adopted and formalized in standards such as ISA-95 (Enterprise-Control System Integration) and ISA-99/IEC 62443 (Cybersecurity for Industrial Automation and Control Systems).

Why are these levels so important when talking to GCs, Building Owners, and Design Teams? Because it mirrors the way they view the building, but in technological terms. It also removes the confusion of our massive, and nerdy, industry. It gives a picture of a whole structure that we are aiming for, but as we move down the model it adds the "plumbing, electrical, airways, walls, and finishes" to make the whole work. When an owner spends $500k on LED retrofits but can't dim them with the AV system in the executive conference room—that's a missed opportunity because there was no Level 3 coordination. It's also one that will be gripped about for years.

The other reason we are moving to this model is that our industry has been challenged for a while with different manufacturers using different but similar terms to market a product. We can't change what vendors call their products. We are going to change how we describe them. If it's in the cloud it's Level 5, think subscription Software as a Service. Corporate network then it's Level 4, think about this as using Active Directory from IT. The Servers that run the building, Level 3. Any controller below the server that has an Ethernet port, Level 2. A controller without ethernet is Level 1. A sensor, actuator, etc. without smarts is Level 0. This is overly simplified, but this is the Purdue model at its most basic.

Now we can start to connect things together. The first thing we can do is make some simple rules. We need systems to work on their own first. The places where they connect have to be very clear.

At the core of this thinking is a concept from software engineering: tight cohesion and loose coupling, popularized by Steve McConnell in his book Code Complete. The idea is simple but powerful: each component of a system should be internally well-designed (tight cohesion), while exposing minimal and standardized points of interaction with other components (loose coupling). In the context of building controls, this means that your HVAC system, lighting system, or access control platform should be robust and fully featured—but not deeply entangled with other systems in ways that make them brittle or hard to update.

Defining internal cohesion and coupling points is what makes a successful integration project. Getting those right makes systems work well on their own but also allow them to share resources with other systems and create interactions greater than what they can do on their own.

It's also the key to future-proofing a facility. A building designed with loose coupling and tight cohesion can adapt. New subsystems can be added. Vendors and systems can be swapped. Cloud platforms can be integrated without disrupting everything else. And systems can evolve on their own terms—without requiring a complete reset every few years.

Throughout this book, I'll guide you in how to:

- Think like an enterprise architect, not just a project manager

- Speak to executives and general contractors about control systems as strategic infrastructure

- Design systems that are modular, standards-based, and ready for change

- Avoid the traps of proprietary systems and hidden integration costs

We are writing this for practitioners who are ready to lead—not just follow specifications. There is a broad base of knowledge needed. The controls industry has matured. Now it's time for our approach to mature with it.

## Executive Summary

### Integrated Control Systems Drive Value
Disconnected HVAC, lighting, security, and AV systems create inefficiencies, overlap, and missed opportunities for comfort, safety, and energy savings.

### The Purdue Model Brings Clarity and Structure
The Purdue Model is a proven, hierarchical framework that aligns building technology with how executives, IT, and facility teams think—making integration strategic, not accidental.

### Start with Strategy, Not Hardware
Successful buildings begin with integration planning from the top down—not just shop drawings and device lists.

### Loose Coupling Future-Proofs Your Investment
Systems built with tight internal design and clear connection points can evolve, scale, and swap vendors without disruption.

### Speak a Common Language Across Teams
The Purdue Model helps executives, general contractors, and tech teams align around a shared vision of how the building's systems should work together.

# Chapter 2: Start at the Top: Enterprise Strategy and Control Integration (Levels 4–5)

*Imagine planning a multi-million-dollar smart building—only to realize, three years later, that your systems can't report on the data your board needs for Environmental, Social, and Governance (ESG) compliance. Why? Because you started at the bottom.*

Before we dive into device networks, protocols, or sequences of operation, we have to start at the top: **strategy**.

Levels 4 and 5 in the Purdue Model represent the **enterprise layer**—the point at which building control systems intersect with business goals, energy policy, digital transformation, and long-term asset value. Historically, building controls were scoped from the bottom up, with each trade independently delivering its piece of the puzzle. But in this new era of smart buildings, cloud integration, ESG reporting, and IT-led capital planning, we must begin at the top.

---

**The "First" Question to Ask on Every Project**

**What is the enterprise strategy?** Really—what does the business want to accomplish?

This will differ in every project. A **core and shell** building will have radically different goals than a **purpose-built,**

**owner-occupied** facility. In a core and shell scenario, the priority may be limited to utility metering or occupancy monitoring, while energy efficiency is considered the tenant's concern—possibly even a revenue stream for the owner.

In contrast, an owner-occupied facility might need to showcase the organization's technological capabilities, or maintain high resiliency to support critical operations. Their enterprise strategy may call for tight controls, layered backup systems, and rich data visibility from day one.

**Understanding these goals early is the only way to design controls that deliver lasting value.**

---

### Vendor-Neutral, Trade-Agnostic Structure

Instead of relying on brand-specific features or siloed contract language, the Purdue Model encourages teams to speak in terms of **function and responsibility—not product**. This makes it easier to:

- Support modularity

- Encourage open integration

- Help stakeholders from different disciplines understand each other

Even if you're representing a specific product line, resist the urge to lead with features. Start with the problem. Define the business objectives. Then come back and match your product offering to the real need. Find the gaps early. You'll save time, money, and a lot of headache later.

This model also aligns directly with how **IT departments** design networks and structure security policies. IT professionals are already fluent in concepts like **network segmentation, layered architecture, and lifecycle planning**. Presenting building controls within this same framework bridges the communication gap between facilities and IT—making collaboration smoother and cybersecurity implementation more consistent.

---

### Bridging the Mindset Gap Between IT and Facilities

This is a hard concept for some to grasp—especially those coming up through the trades—but it's critical to start **soft-selling** the IT mindset from the beginning.

IT teams know:

- There will be **maintenance**

- There will be **support and operations**

- And there will absolutely be **upgrades**

They're buying technology, and they've lived this lifecycle for decades. In contrast, most facilities departments still think in terms of installation and break/fix cycles. Very few have dedicated C-suite representation. But IT does—and **your general contractor is already interfacing with that CIO or CTO.**

If you can get IT on your side, they will often advocate for **long-term support**, **budget allocations**, and even **future-proofing strategies** that make your solution more valuable over time.

---

**Why Start at the Top?**

**Note to Design Teams:**
Starting at Levels 4–5 doesn't just benefit the owner—it makes your job easier. When data requirements and integration expectations are defined early, you can specify systems with confidence, reduce scope gaps, and improve coordination across trades. Control systems should not be defined by whichever trade bids the work first.

A more strategic, top-down approach enables owners to:

- Define **data standards** and **reporting expectations** before installation

- Align operations with organizational goals like **decarbonization** or **resiliency**

- Coordinate **capital upgrades** across systems, not just within one trade

- Avoid **overlapping or incompatible interfaces**

*The higher we start in the design conversation, the more value we can deliver at the lower levels.*

**Levels, Scope, and Audience**

**Level 5 – Portfolio-wide / Cloud Integration**
Audience: Executives, ESG teams, Corporate IT

**Level 4 – Site-Level / Facility IT Systems**
Audience: Engineers, Facilities teams, Security personnel

Here are some examples these Level 4 and 5 services at the intersection of physical infrastructure with **enterprise-level business systems** and **cloud-enabled services** broken down by level, all focused on how OT data and functionality connects with business value:

---

**Level 5 – Enterprise / External Business Services**

These are typically **cloud-based**, **portfolio-wide**, or **corporate IT-managed** systems. They use data from the building to inform decisions at the **organization level**, not just site operations.

**Examples:**

- **Energy Management Platforms**
  e.g., Schneider EcoStruxure, Siemens Desigo CC Cloud, GridPoint

Aggregate energy data across facilities to optimize usage, report on savings, or support decarbonization goals.

- **ESG Reporting Tools**
  e.g., Measurabl, Envizi, Microsoft Sustainability Manager

Pulls data from BAS and submeters to support ESG disclosures and sustainability KPIs.

- **Digital Twin Platforms**
  e.g., WillowTwin, Autodesk Tandem

Mirror building operations in a virtual model using real-time OT data to inform maintenance, analytics, or investment planning.

- **Enterprise CMMS with API Integration**
  e.g., IBM Maximo, HxGN EAM, Archibus, openMaint

Connects with building systems for automated fault detection and work order generation.

- **Corporate Dashboards**
  e.g., Power BI, Tableau, custom dashboards

Executive-level summaries of facility metrics: energy use, comfort conditions, occupancy, risk flags.

- **Utility Program Interfaces**
  e.g., Demand response, curtailment services, time-of-use optimization

Communicate directly with utilities or aggregators using OT data and controls to shift loads or earn incentives. *Author's Note: Some older versions of these demand responses are actually a Level 0 in the BAS perspective, and some are not even automated. This is acknowledged but not the types that are being discussed.*

---

## Level 4 – Site-Level Business Services / Facility IT

These systems are typically **on-site or hosted privately**, used by **facilities managers, building engineers, and IT staff**. They make decisions based on localized data but are often designed to support strategic goals.

**Examples:**

- **Building Management Dashboards**

Single-pane-of-glass views showing HVAC, lighting, access, and alarms across the site.

- **Analytics Engines and FDD (Fault Detection & Diagnostics)**
  e.g., SkySpark, CopperTree, BuildingFit

Analyze BAS data to identify inefficiencies, faults, and trends.

- **Submetering and Utility Billing Platforms**
  e.g., Lucid BuildingOS, eSight

Track energy and water use for tenant billing, internal cost allocation, or conservation efforts.

- **BIM + FM Integration Tools**
  e.g., Autodesk Forge-based viewers

Use 3D building data to locate equipment and guide technicians in the field.

- **Authentication, Data Storage, Time Services, Remote Access** e.g., Active Directory, Database Storage, or other corporate IT services

These are the interfaces to the corporate IT services.

---

**In Summary:**

**Levels, Focus, and Example Use**

**Level 5 – Cloud and Corporate Integration**
Focus: Cloud-based and corporate-wide platforms
Example Use: ESG compliance, portfolio analytics, enterprise asset tracking

**Level 4 – Site-Level Business Systems**
Focus: On-site facility and operational systems
Example Use: Fault detection, energy dashboards, tenant billing, security system integration

**Case Study : Failed Smart Energy Management in Phoenix**

*Background*: A Platinum LEED-certified office tower in Phoenix (McHugh, 2025) implemented an advanced energy management system aimed at optimizing efficiency. However, the project faced significant challenges due to a lack of strategic alignment with the existing urban infrastructure.

The building's systems were not effectively integrated with the outdated local power grid, hindering participation in demand response programs. Efforts to incorporate renewable energy sources were constrained by the grid's capacity limitations. The building was unable to capitalize on real-time energy pricing advantages due to these integration issues.

*Lesson Learned:* This case underscores the necessity of aligning building-level energy strategies with the broader capabilities and limitations of municipal infrastructure. A comprehensive, strategy-first approach could have identified these constraints early, allowing for more effective planning and integration.

Now that we've aligned with enterprise goals, it's time to ensure that every layer below—down to the physical devices—delivers on that promise. In the next chapter, we'll begin unpacking how OT systems fit into this broader strategy.

## Executive Summary

### Start with Strategy, Not Devices

Most building control systems are still designed from the bottom up, but this approach often leads to costly gaps in reporting, integration, and long-term value. By starting at the enterprise level (Purdue Levels 4–5), stakeholders can align building systems with ESG goals, digital transformation, and lifecycle planning from the outset.

### Bridge Facilities and IT Through a Shared Model

Using the Purdue Model helps break down trade silos and align control system planning with IT practices—like segmentation, lifecycle support, and cybersecurity. Speaking in terms of function and business value (rather than product features) creates a shared language between facilities, IT, and executives.

## Define Business Use Cases Before Engineering the System

Enterprise-level tools like ESG reporting platforms, analytics engines, CMMS integrations, and dashboards all rely on clear, early definitions of what data is needed and why. Integrators who ask the right Level 4–5 questions can design more adaptable, modular, and future-ready systems.

# Explainer: What is PropTech?

**PropTech** (short for *Property Technology*) is the umbrella term for the wave of digital technologies transforming how people **buy, sell, rent, manage, design, and operate real estate**. Think of it as the real estate industry's version of FinTech — but for buildings instead of banks.

---

## In a nutshell:

PropTech is where **property meets innovation**, spanning everything from apps that help you find an apartment to AI-driven systems managing entire building portfolios.

---

## Categories of PropTech

### Smart Buildings and Building Automation
HVAC, lighting, security, and energy systems controlled by sensors, software, and integrations.
This book focuses primarily on the Smart Buildings and Building Automation category. While we may touch on other categories, this is where our primary discussion lives.

### Real Estate Platforms
Zillow, Redfin, CoStar, and similar platforms for property search, valuation, and transactions.

### Construction Tech (ConTech)
BIM, drones, prefab automation, AR/VR site planning, and digital twin technologies.

### Property Management Tech
Tools for leasing, maintenance requests, rent collection, and tenant engagement.

### Investment and Finance Tech
Crowdfunding, digital mortgage platforms, fractional ownership, and blockchain for property titles.

### Sustainability and ESG Tools
Carbon tracking, water use analytics, and systems that help buildings meet LEED, WELL, or ESG benchmarks.

---

**PropTech professionals—primarily from IT, finance, and real estate—operate almost exclusively at Level 5 of the Purdue Model.** They focus on portfolios of hundreds of buildings, not individual sites, and their daily concerns revolve around performance metrics, financial models, and data-driven decisions.

Operational Technology (OT) plays a critical role in enabling those outcomes, but it's not part of their everyday workflow. From their perspective, OT systems are expected to function reliably in the background and deliver clean, usable data to the platforms and dashboards they depend on.

**They don't need to know how it all works—just that it does.**

## PropTech and Level 5 Disconnect

PropTech companies—whether optimizing energy use, tracking ESG performance, or creating digital twins—operate predominantly at Level 5 of the Purdue Model. From their vantage point, buildings are data generators and financial assets. But here's the rub: they assume the layers below are already reliable and integrated.

The reality on the ground is often messier. Systems at Levels 0 to 3 are siloed, misaligned, or simply underperforming. Sensors drift. Schedules conflict. Controllers fall out of sync. Yet PropTech dashboards keep updating—based on incomplete or misleading data.

If Level 5 wants to generate real value, it must partner with integration professionals working from Level 3 downward. That's where true system health lives. Without that alignment, your ESG report might just be a glossy abstraction.

Bottom Line: The "future of buildings" needs to be built on solid integration fundamentals. Level 5 can't afford to ignore the messy middle.

# Chapter 3: Integration Checklist for Controls Professionals – Levels 4–5

**Purpose**

Identify business drivers, IT alignment, data requirements, and strategic initiatives before diving into design or product selection.

---

## Enterprise Strategy and Business Objectives (Level 5)

### Ask These Questions:

- What are the top business goals for this facility or portfolio?
  (For example: ESG compliance, energy savings, tenant experience, uptime, decarbonization.)

- Is this a core-and-shell project, tenant improvement, or owner-occupied facility?

- Do you have enterprise-level ESG or sustainability reporting requirements?
  (Are there specific KPIs or frameworks such as GRESB, CDP, LEED, or Net Zero?)

- Are there any corporate-mandated platforms or dashboards we must integrate with?

- What kinds of executive-level reporting or analytics are expected?
  (How often? Who are the stakeholders?)

- Will this site be part of a broader digital twin or smart campus initiative?

- Are you pursuing demand response, curtailment, or grid-interactive capabilities?

---

## Corporate IT, Integration, and Lifecycle Planning (Level 5 and Level 4 Bridge)

### Ask These Questions:

- Has IT been involved in planning for building systems or devices?

- Who is responsible for long-term support, security, and upgrades?

- Will these systems connect to the corporate network or remain air-gapped?

- What are the cybersecurity requirements? (For example: NIST 800-53, ISO 27001.)

- Do you have existing CMMS, EAM, or IWMS platforms?
  (Do you want integration for auto-fault detection or work order creation?)

- Do you want centralized user authentication (such as Active Directory)?

- Will remote access be required? If so, how should it be secured?

## Facility Operations and Site-Level IT (Level 4)

**Ask These Questions:**

- Who will operate and maintain the systems day-to-day?

- Are you using any building analytics or FDD platforms? If not, would you like to?

- Do you require tenant-level submetering or billing integration?

- Is there a single-pane-of-glass dashboard for building systems? Do you want one?

- What tools do your field technicians use today? (For example: Mobile apps, BIM viewers.)

- How is system data currently archived, backed up, and accessed?

- Is there an on-site data historian, or are you expecting cloud-based storage?

- Are there requirements for timestamp accuracy or time synchronization?

- Are building systems part of an internal risk register or critical infrastructure list?

## Data, Visualization, and Interoperability (Level 4–5)

**Ask These Questions:**

- What data points are most important to you or your stakeholders?
  (For example: Energy, indoor air quality, uptime, space utilization.)

- Are there standard naming conventions or tagging schemes? (Such as Project Haystack.)

- Do you want to export data to Power BI, Tableau, or another BI platform?

- How long should data be stored? Who needs access? How frequently?

- Do you want to visualize data in a digital twin or 3D environment?

- Are any systems required to push or receive data via APIs?

---

## Change Management, Budgeting, and Future Proofing

**Ask These Questions:**

- Are you planning phased construction or future expansions?

- How often are upgrades expected or budgeted for?

- Is there funding set aside for system maintenance or IT refresh cycles?

- Would you benefit from modular or scalable integration approaches?

- Who signs off on IT or building technology purchases?

- Would a demonstration or pilot integration help move the project forward?

---

**Deliverables Checklist**

Ensure the following are part of your integration scope where applicable:

- Enterprise Strategy Alignment Matrix (mapping systems to business goals)

- Stakeholder Roles and Responsibilities (IT, Facilities, Executives)

- Data Flow Diagrams (Level 4–5 view)

- API and Integration Points List

- Cybersecurity Plan and Compliance Checklist

- Lifecycle Support Plan

- Visualization and Dashboard Requirements

- Naming and Tagging Standards Reference

- Cloud vs. On-Premises Data Architecture Summary

- System-of-Systems Map (especially important for portfolios)

# Chapter 4: Designing for Integration from Day One

The best-integrated buildings are not the result of luck or late-game coordination—they are built on smart decisions made before a single wire is pulled or pipe is laid. Integration, if it is going to succeed, must be designed deliberately and early, before individual systems are scoped or trades are contracted.

This chapter focuses on how to begin integration strategy during conceptual design and preconstruction, before silos are built.

---

### Note on System Levels

Not all systems at Purdue Level 3 are isolated. Many send valuable data upstream to Level 4 platforms. However, their primary function—real-time control and coordination—keeps them classified at Level 3. Be cautious not to reassign systems based solely on data flow.

---

### Integration as a Design Discipline

Historically, integration has been treated as a coordination activity—something addressed in the field after systems are already purchased and installed. This is backward. Integration should be a first-class design discipline, considered at the same time as structural, electrical, and mechanical systems.

Starting early allows you to:

- Select systems with compatible protocols

- Define ownership of shared devices (such as occupancy sensors and relays)

- Ensure cross-discipline sequences are captured

- Allocate network ports and switch capacity

- Budget for shared infrastructure (such as servers, routers, cloud access)

When integration is an afterthought, it becomes a source of delays and finger-pointing. When it is part of the design, it creates value and flexibility.

---

## Early Integration Planning Checklist

Ensure that early integration planning includes:

- Are system protocols and platform requirements coordinated across specifications?

- Has an integration matrix been started and reviewed with trades?

- Are owner IT policies (such as remote access and security) documented?

- Is shared device ownership defined (such as occupancy sensors and meters)?

- Is a common GUI or dashboard strategy discussed?

- Has the responsible party for integration coordination been named?

---

## Integration Ownership Matters

Integration is everyone's concern—but without a clear owner, it falls through the cracks. Assigning a systems integrator, technology consultant, or MEPT lead to own the integration matrix ensures accountability across disciplines. A dedicated master systems integrator (MSI) leading the integration scope is generally the best option.

---

## Working with Architects, Engineers, and Owners

Integration strategy begins with design team alignment. During programming and schematic design, the project team should:

- Identify what systems will be installed and who typically owns each

- Discuss desired user experience—what occupants, engineers, and managers need to see and control

- Set expectations for shared dashboards or unified GUIs

- Define who will operate and maintain the systems post-occupancy

This requires input from architects (for visual and experience concerns), engineers (for system function and energy goals), IT (for infrastructure and security), and the owner (for expectations and lifecycle plans). For a sample checklist of key documents to capture during early design, see Appendix D.

## Integration Starts with Specification

If you want systems to work together, they need to be specified that way. This is where Division 25 (Integrated Automation) can be a powerful tool. Division 25 is a CSI MasterFormat section designed to bring clarity to system coordination—essentially the connective tissue between HVAC, lighting, security, and more.

If included early, Division 25 becomes a powerful anchor point for integration expectations. It creates a space to:

- Define the integration platform, points, and expected interactions

- Specify who is responsible for coordinating protocols and data exchange

- Require BTL-listed devices or adherence to specific open standards

- Include cybersecurity requirements that apply across all systems

Even if Division 25 is not used, integration responsibilities can be spelled out in Division 23 (HVAC), Division 26 (Lighting), or Division 27 (Communications). The key is not to leave it implied or assumed.

## Case in Point

On a recent government building, each trade delivered their own BACnet-capable system—but no one coordinated object naming, point mapping, or time synchronization. Integration required weeks of reprogramming. Had Division 25 requirements and a shared integration matrix been defined during design, these issues could have been avoided.

## The Cross-Trade Integration Matrix

One of the most valuable tools at this stage is the integration matrix: a simple table that maps how systems will interact. For example:

**System: Fire Alarm**
**System: HVAC**
**Interaction Type:** Shut down air handlers
**Notes:** Per NFPA 72

**System: Access Control**
**System: Lighting**
**Interaction Type:** Turn on lights on entry
**Notes:** Lobby and core zones only

**System: AV System**
**System: Shades**
**Interaction Type:** Lower shades during video
**Notes:** AV preset triggers

This matrix becomes a reference point for all design disciplines and contractors and should be updated as the design evolves. See Appendix E for a sample matrix mapping system interactions across trades.

---

**Leveraging Owner Standards and IT Policies**

Institutional and multi-site clients often have:

- Preferred vendors or platforms

- Existing front-end software that new systems must connect to

- Security policies for device access and remote support

- Server requirements (such as VM configurations, operating systems, backups)

It is critical to gather these standards during schematic design, not at submittals. Doing so prevents surprises and ensures systems are designed to plug in from day one.

---

Treating integration as a design discipline is not just about avoiding headaches—it is how you unlock the real potential of smart buildings. By aligning people, specifications, and systems from the start, you do not just build better buildings—you build buildings that get better over time.

In the next chapter, we will explore how to choose the right integration platforms and architecture to make those plans a reality.

---

## Executive Summary

### Designing for Integration from Day One
Smart buildings are not built in the field—they are designed in the conference room. Effective system integration does not happen by accident. It must be intentionally designed during early project phases—before trades are awarded and systems are purchased. Integration is not a coordination task; it is a design discipline that belongs alongside structural, electrical, and mechanical engineering.

### Start Early
Integration strategy should be developed during programming and schematic design, with participation from architects, engineers, IT, and the owner.

### Specify Integration Clearly
Use CSI MasterFormat Division 25 or international equivalents (such as DIN 276 or Uniclass) to define responsibilities, protocols, and platform expectations.

### Appoint an Integration Owner
Without clear ownership—ideally a Master Systems Integrator—integration falls through the cracks and becomes a source of delays.

### Use an Integration Matrix
A simple, evolving table mapping cross-trade interactions helps clarify dependencies and prevent scope gaps.

### Respect Owner Standards
Capture IT policies, platform preferences, and cybersecurity expectations early to avoid costly redesigns later.

### Real-World Consequences
Projects without early integration planning often face reprogramming, delays, and interoperability failures—even when "open" systems are used.

---

### Bottom Line
When integration is embedded in the design from day one, the result is not just fewer problems—but a smarter, more flexible building ready for long-term success.

# Chapter 5: Platform Selection and Architecture Planning

Before trades are selected and bid packages are issued, decision-makers—especially at the executive level—must ask a critical question:

**"What platform is going to tie all this together?"**

This decision sets the tone for integration success and directly impacts the building's long-term adaptability, operational efficiency, and data strategy.

Choosing the right platform—and planning the architecture that supports it—is where strategy becomes structure.

---

### Integration Does Not Equal Centralization

Imagine a centralized system where HVAC, lighting, and access control are all routed through a single vendor's box. It is efficient—until that box fails or the vendor goes out of business.

Now contrast that with an integrated platform where each system operates independently but shares data through a central translator or message bus. You retain modularity, gain resilience, and avoid being locked into a single vendor's ecosystem.

Integration does not mean putting everything on one box. It means ensuring systems can share information, respond to each other, and provide value—without creating a tangled mess of dependencies.

A good integration platform acts more like a switchboard or translator than a monolith. It lets systems be modular while cooperating through "tight cohesion, loose coupling." Each subsystem (lighting, HVAC, access control, etc.) excels independently but communicates clearly with its neighbors.

This does not necessarily mean a single server. It could mean a single physical device hosting multiple virtual machines. HVAC and AV systems, for example, often require different tools. A platform enclave for OT equipment can reduce IT hardware and improve redundancy and failover.

**Note:** Fire Alarm systems are the notable exception. Due to code requirements, they typically must remain on their own physical IT hardware and cabling. These regulations are changing slowly.

---

### Types of Platforms

Integration platforms are not mutually exclusive and are often used together:

### Building Automation System (BAS) Front-End

Typically provided by the HVAC controls vendor. Offers trend logs, alarm management, and scheduling.
Example: A Tridium-based BAS handling VAV controls and zone scheduling.

**Enterprise or Multi-Site Integration Platform**

Aggregates systems across sites and disciplines, often sitting above BAS systems.

Example: A platform integrating HVAC, lighting, and access control across a university campus.

**Energy Management and Utility Monitoring Platform**

Focuses on metering, utility dashboards, and demand response.

Example: An energy dashboard comparing real-time power usage across multiple buildings.

**IT-Managed Dashboards and Aggregators**

Pulls data from OT systems into cloud dashboards via APIs, BACnet/IP, or MQTT.

Example: A cloud dashboard aggregating data from multiple BAS front ends across retail locations.

Your platform strategy should define:

- Who owns the system

- Where the data resides

- How subsystems connect (protocols and interfaces)

- What happens when systems are replaced or upgraded

- Whether you need one platform or multiple platforms

---

**Cloud vs. On-Premises**

Modern platforms often offer cloud-based features, but not everything should be cloud-based.

**On-Premises Advantages:**

- Real-time responsiveness

- Reduced external dependencies

- Greater control over security

**Cloud Advantages:**

- Easier remote access

- Vendor-managed updates

- Lower initial hardware cost

**Hybrid models** (local core with cloud backup or analytics) are increasingly popular.
Early involvement of IT and cybersecurity teams is critical.

---

## Open vs. Proprietary vs. Framework-Based Platforms

### Open Systems
Definition: Built on open protocols (such as BACnet, Modbus, MQTT) with open APIs.

Advantages:

- Vendor flexibility

- Long-term maintainability and scalability

- Broader community support

Challenges:

- More integration work up front

- Variability in implementations

- Shared responsibility for system performance

Vendor Lock-In Risk: **Low**

*Author's Note:*
While open platforms are ideal, truly open commercial systems are rare. Tools like Node-RED and Sedona exist but are not widely implemented commercially as platforms yet.

---

**Proprietary Systems**
Definition: Vendor-controlled protocols, software, or hardware.

Advantages:

- Turnkey solutions with faster deployment

- Single source of accountability

- Tightly integrated ecosystems

Challenges:

- Long-term dependence on a single vendor

- Limited interoperability and future-proofing

- Upgrade paths controlled by the vendor

Vendor Lock-In Risk: **High**

## Framework-Based Systems

Definition: Platforms that are not fully open but provide structured extensibility.

Examples: Niagara Framework, Microsoft Azure Digital Twins, KNX.

Advantages:

- Balance of flexibility and structure

- Ecosystem of compatible vendors or certified integrators

- Vendor-neutral upper layers

Challenges:

- Framework itself can become the lock-in

- Licensing and certification complexities

- Skilled integrators needed

Vendor Lock-In Risk: **Moderate**

*Appendix* J includes a deeper comparison of Niagara and open frameworks for Division 25 planning.

## Why This Matters

Understanding this spectrum helps stakeholders ask better questions during design:

- Are we building for long-term flexibility or short-term convenience?

- Who owns the data—and how easily can it be moved?

- If something breaks or needs an upgrade, who can fix it?

## Architecture Planning

Once you choose your platform, plan the infrastructure:

- IT network backbone: Converged, standalone, or hybrid

- VLANs, port assignments, and firewall rules

- Redundancy and failover: What happens if the BAS server fails?

- Time synchronization: NTP setup across devices

- Bandwidth planning: Important for IP video and high-volume metering

Approach it like electrical distribution—plan the backbone first.

## Ownership and Responsibility

Post-commissioning failure often happens because no one clearly owns the platform.

To prevent this:

- Assign roles early using a RACI chart (Responsible, Accountable, Consulted, Informed)

- Define access permissions and roles

- Set training and turnover expectations

- Write ownership requirements into the specification

---

**Sample RACI Table: Integration Platform Ownership**

**Define integration strategy:**
Responsible: Consultant
Accountable: Executive
Consulted: Facility Team
Informed: General Contractor

**Select platform architecture:**
Responsible: Consultant
Accountable: IT Department
Consulted: Facility Team
Informed: GC/Owner

**Maintain platform post-turnover:**
Accountable: IT Department
Responsible: Facilities
Consulted: Integrator
Informed: Owner

(Expanded RACI charts are included in Appendix K.)

---

**Executive Summary**

### Integration Requires Early, Intentional Design

True integration is about communication between modular systems—not putting everything on one server. Platform selection before bidding avoids lock-in and supports long-term flexibility.

### Plan Architecture Like Infrastructure

Network topology, failover plans, time synchronization, and bandwidth considerations are foundational for successful integration.

### Understand Your Platform Spectrum

Open, proprietary, and framework-based platforms each offer different trade-offs. Choosing the right fit requires understanding both current and future needs.

### Assign Ownership and Document It

A clear RACI matrix defining operational responsibility prevents systems from falling into disrepair after turnover.

---

### Bottom Line

Integration begins with intention—and that intention must be supported by a resilient, well-structured, and clearly owned platform strategy.

# Chapter 6: Cybersecurity and System Governance

No one ever thinks the BAS is going to be the weak link—until it is.

*"Seventy percent of OT cyber incidents are due to poor security hygiene and unpatched systems."* (Dragos, Inc.)

Whether it's a ransomware attack, a misconfigured switch that exposes building devices to the internet, or a technician who reuses passwords across client sites, building systems are increasingly in the crosshairs of cybersecurity threats.
Unlike traditional IT systems, building controls often fall between trades, vendors, and departments.

This chapter lays out a path to secure, supportable building automation through smart policies, realistic planning, and shared responsibility.

---

### Start Early: Build Security into Design

If you wait until the building is operational to think about cybersecurity, you are already behind.
Security must be designed into the system architecture—starting with vendor selection and platform design.

Key early actions:

- Define cybersecurity responsibilities in Division 25 or owner standards.

- Require vendors to comply with a known framework such as NIST SP 800-82 or IEC 62443.

- Engage the owner's IT team in architecture review.

- Verify firewalls, VLANs, and authentication requirements during design—not during commissioning.

Security isn't just a punch list item; it's an ongoing obligation.

---

**Standards to Know: NIST, IEC, and Real-World Practice**

There are two gold standards for OT cybersecurity:

- **NIST SP 800-82** – A U.S.-focused guide for industrial control system security.

- **IEC 62443** – A global standard for securing industrial automation and control systems.

Both emphasize:

- Segmentation between IT and OT networks.

- Role-based access control.

- Secure remote access methods, including VPNs and multifactor authentication.

- Logging, patching, and lifecycle policies.

The key takeaway: You don't have to invent a new approach—you just need to select one and implement it thoroughly.

---

## Policies that Matter (and Can Be Enforced)

Cybersecurity policies must be simple enough to be understood and enforced in real-world operations.

Focus on the basics:

- **Access Control**: No shared logins. Require named accounts and multifactor authentication for remote access.

- **Password Policies**: Enforce strong passwords with regular updates. Store credentials securely.

- **Network Segmentation**: Place OT systems on isolated VLANs with restricted cross-traffic through firewall rules.

- **Remote Access**: Use secured remote access methods (VPN, jump box, or managed platform). No open inbound ports.

- **Patching and Updates**: Schedule periodic patching windows. Avoid "set it and forget it" systems.

- **Audit Logging**: Enable change logging and archive logs for both diagnostics and forensics.

These policies only work if they are assigned to someone for maintenance and oversight.

## Cybersecurity and System Governance

In building automation cybersecurity, confusion often arises around one key question:
**Who is responsible for keeping systems secure?**

The typical assumptions:

- Vendors deliver the equipment and baseline configurations.

- Integrators install and sometimes harden the systems.

- IT departments manage enterprise networks but have little visibility into building control systems.

- Owners assume long-term risk but often lack visibility into OT vulnerabilities.

A better framework:

- **Vendor**: Responsible for delivering a secure, updateable product.

- **Integrator**: Accountable for the secure implementation of that product.

- **Owner**: Accountable for ongoing governance, updates, and monitoring.

This triad must be defined clearly in contracts, scopes, and specifications.

As I argued in my white paper, *Cybersecurity in Automation – Legality and Ethics*, handing off a vulnerable system to a client isn't just sloppy—it may be legally negligent.

---

## Who Owns This?

Cybersecurity doesn't work when it's treated as "someone else's problem."
Ownership models that work include:

- **IT-Owned Security**: IT defines the rules, and vendors implement them.

- **Shared Governance**: Facilities manage operations; IT manages infrastructure and monitoring.

- **Third-Party Managed Services**: A trusted integrator maintains updates, credentials, and audit logs.

Whatever model is chosen, it must be documented before project turnover.

---

## Writing It Into the Spec

If cybersecurity requirements are not in the specification, they will not exist in practice.

Include the following:

- Required frameworks (such as NIST SP 800-82 or IEC 62443).

- Minimum password and authentication standards.

- Consider mandating a password vault for credential storage.

- Secure protocol and port policies.

- Remote access methods and restrictions.

- Vendor support obligations (such as patching SLAs and credential handling).

- Turnover documentation requirements, including admin credentials, network diagrams, and firewall rules.

Cybersecurity language can reside in Division 25, Division 27, or Division 01—best practice is 25 05 11 – Cybersecurity Requirements for Facility Controls.

---

**Case Study: Google Australia's Wharf 7 Office**

In 2013, researchers Billy Rios and Terry McCorkle discovered major vulnerabilities in Google's Wharf 7 Building Management System (BMS), based on the Niagara AX platform. (Zetter, 2013)

Findings:

- Administrative password: "anyonesguess."

- Outdated security patches.

- Unauthorized access to HVAC control panels.

Lessons learned:

- Patch systems promptly.

- Use strong, unique passwords.

- Isolate critical systems from the public internet.

Google responded quickly, disconnecting the system from the internet and reinforcing internal protocols.

---

## Case Study: Target Corporation Vendor Breach

In 2013, attackers gained access to Target's network through credentials stolen from an HVAC vendor, Fazio Mechanical Services. (Cupertino Electric Inc., 2015)R

Process:

- Phishing email led to credential theft.

- Attackers accessed Target's internal network.

- Malware installed on POS systems, stealing customer payment data.

Lessons learned:

- Strict cybersecurity requirements for third-party vendors are critical.

- Proper network segmentation could have limited attacker movement inside the network.

---

## Case Study: Oldsmar, Florida Water Treatment Plant

In February 2021, an attacker remotely accessed the Oldsmar water treatment facility and attempted to dangerously raise the sodium hydroxide concentration.

Findings:

- Weak or reused credentials.

- Insecure remote access (TeamViewer without multifactor authentication).

- Outdated Windows 7 system.

Response:

- An alert plant operator caught the unauthorized activity live.

Lessons learned:

- Legacy systems and shared credentials create serious vulnerabilities.

- Multifactor authentication and network segmentation are essential.

---

## Case Study: Government Field Office Incident

In 2014, a controls vendor used TeamViewer installed on client servers—without disclosing the practice—to provide remote support, including at a government facility.

Findings:

- Stored credentials on multiple client systems.

- No disclosure to clients.

- No multifactor authentication.

Impact:

- Although no breach occurred, the vendor lost several major accounts and was barred from future government work.

Lesson:
Transparent remote access policies and robust credential security are non-negotiable.

---

**The Goal: Secure, Sustainable Systems**

Perfect cybersecurity is unattainable.
**But practical, repeatable, and supportable cybersecurity is achievable.**

A strong cybersecurity posture:

- Supports system uptime and reliability.

- Protects brand reputation.

- Ensures systems age gracefully alongside IT infrastructure.

Cybersecurity must be seen as part of system operations—not a last-minute add-on.

---

## Executive Summary

**Key Idea:**
Cybersecurity for building systems isn't optional. It must be designed in, maintained over time, and governed through real, documented roles and policies.

**Why It Matters:**
BAS, SCADA platforms, and other OT systems are increasingly targeted through weak passwords, default configurations, insecure remote access, and unpatched vulnerabilities.

**Action Steps:**

- **Start at design:** Define cybersecurity responsibilities, frameworks (NIST SP 800-82, IEC 62443), and secure architectures early.

- **Establish enforceable policies:** Require strong passwords, secure remote access, and audit logging. Use credential vaults.

- **Define governance:** Assign clear cybersecurity ownership—IT, facilities, or managed services—and document it.

- **Write it into specs:** Include cybersecurity requirements clearly in Division 25, Division 27, or Division 01.

- **Plan for operations:** Empower facilities teams with checklists and response plans.

**Real-World Validation:**

From Google's Wharf 7 BMS exposure, to Target's vendor credential breach, to Oldsmar's SCADA attack, real-world cases show why proactive cybersecurity is critical.

**The Goal:**

Not perfection—but resilient, sustainable, and supportable protection for building systems.

# Sidebar: Vendor Lock-In as a Cybersecurity Risk

Vendor lock-in is often treated as a procurement or operational inconvenience, but in the context of operational technology (OT), it poses a significant cybersecurity risk. When an organization becomes overly dependent on a single vendor for their systems, tools, and support, they inherently surrender a degree of control over their security posture. Proprietary platforms often limit access to logs, configurations, and even firmware—hindering transparency and making it difficult to independently verify system integrity or respond effectively to threats.

This lack of visibility becomes especially dangerous when responding to cyber incidents. OT teams may find themselves unable to isolate devices, apply patches, or deploy custom security controls without vendor intervention. This delays containment and remediation efforts, giving adversaries more time to exploit vulnerabilities or move laterally within the network. In some cases, the vendor may not even be able to respond quickly—especially if they're overseas, understaffed, or have sunset the product.

Proprietary protocols and closed ecosystems also block integration with modern cybersecurity tools. This stifles innovation and prevents organizations from deploying best-in-class detection and response technologies. Additionally, many legacy OT systems locked to a vendor reach a point where updates or security patches are no longer available. At that point, critical infrastructure becomes reliant on unsupported, unpatchable equipment—essentially hard-coding a vulnerability into the system.

Beyond technical concerns, vendor lock-in introduces strategic and compliance challenges. Organizations may struggle to meet evolving regulatory requirements for logging, access control, or auditability if the vendor's system can't support those features. It also limits the organization's ability to modernize its architecture—for example, by adopting zero-trust principles or microsegmentation—because core systems are only compatible with the vendor's legacy assumptions.

In a top-down integration strategy, avoiding or mitigating vendor lock-in is not just about preserving future flexibility—it's about maintaining security, resilience, and control. Decision-makers must treat vendor selection and ecosystem design as part of the cybersecurity strategy, not an afterthought.

## Explainer: What Is a Password Vault?

A **password vault** is a **secure, encrypted repository** used to store, manage, and control access to sensitive credentials—like usernames, passwords, SSH keys, and API tokens. Think of it as a **secure credential manager** with features like:

- Encrypted storage
- Audit logging
- Role-based access control (RBAC)
- Password rotation
- Session recording (in some systems)
- Integration with Active Directory or LDAP

Popular enterprise-grade options include:

- **CyberArk**
- **HashiCorp Vault**
- **Keeper**
- **BeyondTrust**
- **Thycotic / Delinea**
- For smaller use cases: **Bitwarden**, **LastPass (Enterprise)**, or **KeePassXC (offline)**

---

**Why Use a Password Vault in OT Projects?**

Unlike IT systems that are often maintained by full-time sysadmins, **OT environments** (BAS, SCADA, PLCs, etc.) are:

- Designed by **multiple vendors**

- Managed by **facilities personnel or integrators**

- Expected to last **10–30 years**

- Frequently forgotten in **cybersecurity planning**

**Credential Vault Features and Benefits for OT Projects**

**Feature:** Centralized Storage
**Benefit:** Store PLC, BAS, router, and switch credentials in one place.

---

**Feature:** Controlled Access
**Benefit:** Limit who can view or edit passwords (for example, HVAC vendors cannot access security camera credentials).

---

**Feature:** Audit Logging
**Benefit:** Track when and who accessed or changed a password.

---

**Feature:** Credential Lifecycle
**Benefit:** Rotate passwords periodically or when contractors or staff change.

---

**Feature:** Emergency Access
**Benefit:** Provide trusted personnel with controlled access during off-hours.

---

**Feature:** Remote Access Integration
**Benefit:** Some vaults support SSH or RDP tunneling for secure remote maintenance.

---

## Common OT Risks Without a Vault

- **Hardcoded passwords** (e.g., admin123) used across multiple controllers

- **Shared logins** among vendors or technicians

- **No audit trail** of who logged in or changed configs

- **Lost credentials** after turnover or system lockouts

- **Delayed response during emergencies** if only one person has the password

---

## Sample Use Case

### Scenario: BAS + Security + Lighting Integration

An integrator sets up fire alarm-triggered lighting and HVAC sequences.
Credentials for:

- BAS front end

- Lighting controller (via API)

- Fire alarm system (read-only BACnet device)
  ...are stored in a secure vault accessible only to the facility engineer and approved integrators. Passwords are rotated annually or when a contractor is offboarded.

---

**Best Practices for OT Vaulting**

1. **Document vault access during commissioning**
   Add it to the Division 25 closeout checklist and password handover docs.

2. **Use vaults with API integrations**
   Vaults like HashiCorp Vault can inject credentials into PLCs or Docker containers dynamically.

3. **Align with NIST / ISA-62443 controls**
   Vaulting satisfies portions of **Access Control (AC)** and **Audit & Accountability (AU)** requirements.

4. **Use local/offline vaults for air-gapped sites**
   Tools like **KeePassXC** can be deployed in offline modes with encrypted backups.

5. **Train facility staff to use the vault**
   Don't just hand over credentials in a spreadsheet—train the long-term operators.

# Cybersecurity Playbook for Building Controls

**Goal:** Implement practical, consistent, and sustainable cybersecurity practices for building systems.

## 1. Segment the Network

- Place OT systems on dedicated VLANs
- Block unnecessary traffic between IT and OT

## 2. Secure Access

- Require named user accounts and role-based access
- Use multifactor authentication for remote access
- Enforce strong passwords and credential expiration

## 3. Manage Vendors

- Require vendors to follow your policy
- Disallow shared logins or hardcoded passwords

## 4. Define Remote Access

- Use VPN or secure portal
- Audit all remote sessions

## 5. Keep Systems Updated

- Schedule patch windows every 6–12 months
- Track software and firmware versions

## 6. Monitor and Log

- Log user activity and system changes
- Store logs off-device when possible

## 7. Assign Ownership

- Document roles and responsibilities
- Decide who is accountable for ongoing cybersecurity

## 8. Include in the Contract

- Put cybersecurity expectations into the spec and scope
- Require turnover documentation including user access, firewall rules, and patch logs

Remember: You don't have to be bulletproof—but you do need to be better than "default settings."

# Cybersecurity Design & Implementation Checklist for Building Control Systems

This checklist helps ensure that essential cybersecurity measures are addressed at each project phase.

---

## Design Phase

- o **Cybersecurity framework selected** (e.g., NIST SP 800-82 or IEC 62443)
- o **Defined OT network segmentation** and VLAN structure in network drawings
- o **IT department engaged** in control system planning and reviews
- o **Cybersecurity roles assigned** in project org chart or RACI matrix
- o **Security expectations written into specifications** (Div 25, 27, or 01)

---

## Pre-Installation

- o **Vendor compliance confirmed** for all cybersecurity requirements
- o **Approved secure remote access method defined** (VPN, jump host, etc.)
- o **Password policy established** (complexity, expiration, no default logins)
- o **Role-based user accounts** established (no shared accounts)
- o **Firewall and port policies documented** by IT and integrator

---

## Commissioning / Turnover

- ○ **System audit performed** before final acceptance (network scan, port usage, password review)
- ○ **All admin credentials documented and turned over** securely
- ○ **Audit logging enabled** and accessible to facilities or IT
- ○ **Time synchronization implemented** across all control devices
- ○ **Backup and recovery plan delivered** for controllers and servers

## Ongoing Operations

- ○ **Patch management plan in place** with responsibilities and schedule
- ○ **User access review policy established** (quarterly or bi-annually)
- ○ **Remote access session logging verified**
- ○ **Change control process defined** for system updates or reconfiguration
- ○ **Incident response procedure documented** (who to call, what to do)

Use this checklist on every controls project—whether it's a standalone BAS or a fully integrated building. Cybersecurity is a team sport, and clarity keeps you secure.

# Explainer: Simplified Cybersecurity Incident Response Plan (IRP) for Operational Technology (OT) environment.

**Organization:** [Your Facility or Utility Name]
**Version:** 1.0
**Date:** [Insert Date]
**Owner:** Facilities Manager / OT or Automation Systems Lead

---

## 1. Purpose

This plan provides clear steps for Facilities and Operations staff to recognize, report, and respond to cybersecurity incidents affecting OT systems like HVAC controls, SCADA, PLCs, access control, and energy management systems.

---

## 2. Who Uses This Plan

This plan is intended for:

- Facilities Technicians

- Controls Engineers

- Plant Operators

- Maintenance Leads

- Supervisors

---

## 3. What to Look Out For (Common OT Cybersecurity Events)

If you have completed equipment troubleshooting procedures and still cannot find a cause, consider that this could be a cybersecurity incident. Watch for these signs:

- Unexpected equipment shutdowns or erratic behavior

- Controllers going offline or rebooting without explanation

- Strange logins or remote access sessions you didn't initiate

- Alarms or messages from the automation system that don't make sense

- Locked-out screens or files (ransomware)

- Unfamiliar devices appearing on your network

---

## 4. What To Do When You See Something

### Step 1: Notify Your Supervisor Immediately

Tell your direct supervisor or lead technician what you observed.
If email or system tools are down, use a phone call.

Say:
"I saw [describe issue] on [system name]. This may be a cybersecurity problem."

If your team concurs that this is not an equipment issue, or you cannot contact your team, proceed to Step 2.

---

### Step 2: Call the OT Cyber Response Contact

Contact your designated OT Cybersecurity lead immediately.
Here is the information to use:

- **Primary Contact:** [Jane Smith]

- **Role:** OT Cybersecurity Contact (Internal or External)

- **Phone/Text:** (xxx) xxx-xxxx

This person is responsible for:

- Assessing the threat

- Calling in outside help if needed

- Coordinating with IT and cybersecurity partners

---

### Step 3: Contain If You Can (Safely)

**If you are trained and it is safe to do so:**

- Disconnect the affected device from the network (unplug the network cable or disable the port)

- Lock physical access to the affected panel or room

- Take pictures of any screens or suspicious messages

- Place systems in manual mode and continue operations without the automation system

**Do NOT:**

- Turn off, restart, or wipe the device

- Share screenshots on personal phones or social media

- Touch systems you are not trained on

- Override or acknowledge alarms or messages on the SCADA system (silencing alarms is acceptable)

---

## 5. What Happens Next (Outside Team Steps In)

The OT Cyber Response Contact will:

- Validate whether an incident occurred

- Engage outside cybersecurity professionals or vendors if needed

- Coordinate with IT or central cybersecurity groups

- Lead recovery, patching, and investigation

Facilities team members may be asked to:

- Provide access to affected equipment

- Explain what was observed and what actions were taken

- Assist with system recovery efforts

## 6. After the Incident

Following any incident, the Facilities team will participate in a short review to discuss:

- What happened
- What worked and what didn't
- Recommendations for improvements

## 7. Contacts List

**Primary Contacts:**

- **John Doe** — Facilities Supervisor — (xxx) xxx-xxxx — First point of contact
- **Jane Smith** — OT Cybersecurity Contact — (xxx) xxx-xxxx — Escalation contact
- **Vendor Name** — BAS/SCADA Integrator — (xxx) xxx-xxxx — 24/7 Support
- **IT Contact** — Cybersecurity Partner — (xxx) xxx-xxxx — For advanced threat handling

# PART II: Applied Systems and Trade-Level Coordination

# Chapter 7: Level 3 – Site and Facility-Level Supervisory Control

If there's one place where integration efforts either come together or fall apart, it's here—**Level 3 of the Purdue Model**. This is the layer where systems are visualized, alarms are handled, schedules are coordinated, and operators interact with the building's technology day in and day out.

But in our industry, Level 3 is also where **the language breaks down**. One vendor calls it the BAS. Another says *front-end*. Someone else refers to a *supervisory workstation*, while others toss around *middleware, head-end*, or simply point at a screen and say, "that thing."

This chapter aims to **clear up the confusion** and standardize what we mean when we talk about the layer that matters most for **day-to-day operations**.

---

**Why Level 3?**

By identifying this tier as **Level 3—following the Purdue Model**—we start to establish a **shared vocabulary** for supervisory systems, regardless of vendor or discipline. It lets us say:

"This is the Level 3 system. It communicates with field controllers (Level 2), aggregates data, exposes the GUI, and hands off to IT or cloud systems at Levels 4 and 5."

When everyone—HVAC, lighting, security, AV, and IT—speaks the same language, it's easier to **coordinate responsibilities**, **define expectations**, and **design integrations that work**.

---

**What Happens at Level 3?**

Level 3 is the **brain center** for visibility, coordination, and interaction. It typically includes:

- **Graphical User Interfaces (GUIs)** – Dashboards, floorplans, system navigation

- **Alarm Management** – Notification rules, suppression logic, escalation paths

- **Trend Logging** – Historical data for performance tracking and diagnostics

- **Scheduling** – Coordinated, time-based operations across subsystems

- **User Management** – Authentication, permissions, and role-based access

- **Protocol Routing / Translation** – Gateway functions between otherwise siloed systems

Level 3 isn't just where systems talk to each other—it's where people interact with the building. Operators, engineers, and technicians rely on this tier every day to make sense of what's happening, why it's happening, and what to do about it.

Level 3 may be delivered as a **single on-prem server**, a **distributed architecture**, or even a **cloud-hosted platform**—but the functional role remains the same: **aggregate and coordinate**.

---

## Names You Might Hear (And What They Really Mean)

Instead of arguing over what to call it, this book simply says: **call it Level 3.**

That said, it helps to know the different terms vendors and trades might use when referring to this layer. For a full breakdown of common terminology—including what each one typically means—see **Appendix A: Naming Conventions Across Systems.**

---

## Coordinating Across Systems

Level 3 is **where the magic happens**—the place where siloed systems come together for intelligent behavior:

- A **fire alarm** triggers shutdown of AHUs (HVAC), unlocks access doors, turns on emergency lighting, and lowers AV volume for clear announcements.

- A **conference room booking** adjusts AV presets, lighting scenes, and temperature before the meeting starts.

- A **holiday schedule** disables lighting and HVAC in unoccupied areas across multiple buildings.

To do this, the Level 3 system must:

- Communicate with multiple **Level 2 devices across trades**

- Support **open protocols** like BACnet, Modbus, MQTT, and REST

- Normalize data for usability (point names, units, states, and remove vendor-specific quirks)

- Handle **priority overrides**, **manual modes**, and **exceptions gracefully**

---

**Breaking the Silos at Level 3**

In many buildings, technologies evolve in silos. HVAC might get a new DDC system during a retrofit. AV could be upgraded independently with its own touch panel and scheduling system. Access control, lighting, and metering may follow suit—each selected, installed, and operated in isolation.

These systems often build up **parallel tech stacks** that operate fine on their own—but don't talk to each other. Each has:

- Its own controllers (Level 2)

- Its own user interface or scheduling logic

- Its own terminology, protocols, and points of contact

What's missing is a **shared context**—and that's exactly what Level 3 provides.

Level 3 is where we can **cross-pollinate data** between disciplines **without requiring a full rip-and-replace**. For example:

- **AV + HVAC**: Room reservations in an AV system can trigger pre-cooling or lighting adjustments by sharing scheduling data through the Level 3 platform.

- **Lighting + Security**: Motion sensors used for lighting control can double as occupancy indicators for access control or HVAC zoning logic.

- **Fire Alarm + Everything**: A fire event at Level 3 can cascade across HVAC, access, lighting, and AV for coordinated emergency response.

By aggregating data and coordinating logic at Level 3, you don't have to rebuild every subsystem—you just need to make them *interoperable* at the supervisory layer.

**The key to integration isn't just protocol compatibility— it's having a shared platform at Level 3 that can translate, normalize, and coordinate.**

During commissioning, Level 3 provides a shared interface to validate cross-system logic, making it easier to test sequences that span HVAC, lighting, AV, and security. Remember this, we are going to come back to it later.

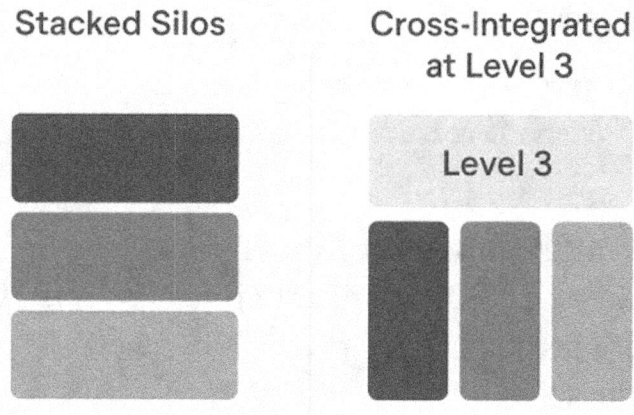

*Figure 1 - Level 3 allows siloing for function, but should feed up for Integration*

---

**IT and Cybersecurity Concerns**

Level 3 systems are often **on shared networks** and may interface with the cloud—making them **high-value targets** for cybersecurity threats.

Design with the following in mind:

- Secure GUI access behind a **firewall or VPN**

- Implement **role-based access control (RBAC)**

- Keep server OS, databases, and libraries **patched and current**

- Monitor logs for **unauthorized access** or **configuration changes**

This is also the right place to implement **network segmentation** and **NTP synchronization**. If timestamps drift, your logs, alarms, and trends lose meaning.

For more on how to design this securely, see Chapter 5 on Cybersecurity and System Governance.

---

**Who Owns Level 3?**

Ownership of Level 3 is often **unclear—and that's a problem**.

- If it's owned by **HVAC**, it may lack visibility into lighting, AV, or access control.

- If it's owned by an **integration contractor**, it requires **buy-in from all trades**.

- If **IT manages it**, there must be proper documentation, backup, and change control.

**Clearly defining ownership during design** ensures this crucial layer doesn't fall through the cracks after handover.

---

**Design Considerations**

When specifying or evaluating a Level 3 system, ask:

- Will this platform support **multiple trades**, or is it HVAC-only?

- Can it **aggregate across buildings or campuses**?

- Is it **user-friendly** with customizable graphics and workflows?

- Does it support **API access** or integration for future capabilities?

- Can it be **documented, maintained, and upgraded** as infrastructure?

In other words, **treat it like core infrastructure**, not a bolt-on widget.

When Level 3 is designed well, everyone wins—operators have insight, IT has governance, trades have coordination, and the building finally starts acting smart.

---

**In the next chapter, we'll drop down to Level 2**—the zone-level controllers and I/O that turn supervisory intent into real-world control.

**Executive Summary**

**Level 3 is Critical for Smart Building Operations**
It is the supervisory control layer where operators interact with building systems, alarms are managed, and cross-trade coordination happens—yet it is often misnamed or misunderstood.

**Integration Lives Here**
Level 3 connects siloed systems (HVAC, lighting, AV, security) through protocol translation and shared logic, enabling coordinated behavior without full subsystem replacement.

**Clear Ownership and Cybersecurity Are Essential**

This layer bridges OT and IT. Ownership must be clearly assigned, cybersecurity integrated from the start, and the system treated as a critical piece of infrastructure.

# Chapter 8: Level 2: Controllers and Local Supervisory Devices

**At Level 2 of the Purdue Model, we leave the boardroom and enter the boiler room. This is where automation logic meets the real world.** From HVAC loops to lighting presets, door control to AV scenes, Level 2 devices are the *local brains* responsible for real-time decisions. But they come with a twist: every trade names them differently—and that's where integration starts to wobble.

Level 2 is where the system comes to life. It's the layer where logic runs, commands are issued, and real-time control takes place. It's the "local brain" of each subsystem, housing the intelligence that interprets sensor data and operates actuators, fixtures, and interfaces.

And it's also where the confusion begins—again. One vendor might call their device a "controller," another might use "processor," "node," or "panel." You might hear about lighting panels, AV processors, programmable logic controllers (PLCs), room controllers, or gateway devices. But in reality, they all serve a similar purpose:

**Level 2 executes control logic and interfaces with Level 1 devices.**

By calling this group Level 2, we give it a neutral, universal label that cuts across silos and product branding.

### Functions at Level 2

Level 2 devices include:

- **DDC controllers** for HVAC

- **Lighting control panels** or distributed relay modules

- **Audio DSPs and AV processors**

- **Access control panels** for security

- **Fire alarm addressable control modules (ACMs)**

- **Custom or hybrid logic controllers** (Raspberry Pi, PLCs, etc.)

Regardless of the trade or label, all Level 2 devices:

- Run local logic

- Manage inputs (like sensors and switches)

- Operate outputs (like motors and relays)

- Talk up to Level 3, or down to Level 1

- Sometimes, act as translators for protocol-specific networks

In some cases, they also serve as gateways to protocol-specific field networks (e.g., DALI, MS/TP, LON).

**Why it gets messy:** Every trade—and even different vendors—use their own labels for devices doing almost the same job. Here's what you might encounter, and what it really means in the context of Level 2.

**Common Terminology (and Mistranslations)**

**Typical Devices and Functions at Level 2**

### HVAC – DDC Controller

Runs sequences, PID loops, and schedules to manage heating, ventilation, and air conditioning operations.

### Lighting – Lighting Control Panel

Manages lighting relays, preset scenes, and scheduling for interior and exterior spaces.

### AV – AV Processor or DSP

Controls audiovisual presets, input routing, mute states, and zone configurations.

### Security – Access Control Panel

Handles door logic, credential verification, and basic access management.

### Fire Alarm – Control Module or Panel

Provides zone monitoring, emergency signaling, and dry contact closures for system coordination.

Instead of debating which system "runs the show," we can define integration at this layer with neutral terminology.

### Cross-Trade Coordination

When coordination breaks down at Level 2, trades start stepping on each other's toes—and devices do too. A lot of headaches happen when Level 2 devices are installed without awareness of their neighbors:

- AV installers might pull cables for wall plates without knowing BAS has a wall sensor going there too

- Security panels might trigger lighting, but with no priority logic to prevent fighting with a lighting controller

- HVAC DDC panels might try to override a damper that's already under fire alarm control

Coordination at Level 2 means:

- Agreeing on point ownership (e.g., who controls a relay or overrides a device?)

- Mapping priorities clearly (e.g., fire alarm always wins, AV scenes can be overridden by occupancy sensors)

- Avoiding duplication of sensing devices (e.g., one $CO_2$ sensor should serve both HVAC and lighting)

**Integration Gone Wrong (Common and Real Project Example)**

- Fire alarm system closed an AHU damper that the HVAC system thought it still controlled, causing a duct blow out

- Two wall plates installed back-to-back: one by AV, one by lighting—fighting for wall real estate, and they control the same conference room

- Security panel triggered emergency lighting—but lighting panel didn't prioritize the signal, or turned on the entire floor rather than the alerted zone

**Moral of the Story:** Without coordination at Level 2, your "smart" building turns into a turf war.

**Protocol Handling**

Level 2 is where most protocol mismatches occur:

- One vendor speaks BACnet/IP, another speaks Modbus RTU

- AV gear might rely on serial commands, while lighting uses DALI

- Some systems need translation layers between proprietary APIs and open protocols

**Sidebar:** Not every translation needs to be a box on a wall. Software-based protocol brokers (like Node-RED or Niagara) can often do the job—if the design allows room for them.

An effective integration design includes:

- Protocol translators where needed (hardware or software)

- Standard naming conventions across systems

- A clear data model that maps each device's capabilities to a common structure

**Resilience and Failover**

Many systems can operate autonomously at Level 2, even if Level 3 is down:

- HVAC controllers can maintain temperature control and respond to schedules

- Access panels can still allow card swipes

- AV processors can execute preloaded scenes

Design for:

- Graceful degradation (what keeps working when the network goes down?)

- Local overrides (manual switches, wall stations)

- Persistent memory (settings, timers, and logic that survive a reboot)Level 2 is the "do" layer.

It's where real-time decisions happen, and where coordination pays off—or fails hard. In the next chapter, we'll examine the systems most people associate with building controls: **HVAC**—and how to rethink those systems as part of a larger integration strategy.

Here's a breakdown of common **Level 2 terminology grouped by industry silo**, showing just how fragmented the naming can be:

---

## Common HVAC (Mechanical Controls)

## Terms and Descriptions

### DDC Controller
Direct Digital Controller; runs logic sequences and PID loops for building systems.

### Unit Controller

Controls a specific piece of equipment, such as a VAV box, air handling unit (AHU), or fan coil unit.

### Zone Controller

Provides smaller-scale logic and control for individual spaces or rooms, often managing temperature or lighting locally.

### Field Panel

An enclosure housing input/output modules and logic hardware, often supporting multiple pieces of equipment or systems.

### RTU Controller

A dedicated controller managing rooftop units (RTUs) for HVAC operation, including compressors, fans, and economizers.

### Plant Controller

Supervises central utility plants, managing boilers, chillers, pumps, and related sequences for large-scale heating and cooling.

---

## Common Lighting Control Terms and Descriptions

### Lighting Control Panel

Contains relays, scheduling functions, and sometimes local logic for managing lighting zones.

### Room Controller

Manages lights for a single room or zone, often integrating with occupancy sensors or wall stations.

### Zone Controller

Controls lighting across multiple rooms or larger open areas, coordinating groups of fixtures or zones.

### Digital Switch Pack

Receives commands from sensors or control panels to switch or dim lighting loads remotely.

### Fixture Controller

An integrated controller embedded directly in individual light fixtures, providing localized logic and control.

### Scene Controller

Applies pre-set lighting configurations for group scenes, commonly used in conference rooms or AV environments.

---

## Common Security and Access Control Terms and Descriptions

### Access Control Panel

Centralized controller managing the logic for door readers, locks, and alarm points across a building or site.

### Door Controller

Manages control and monitoring for one or two doors, including reader inputs and strike relays.

### Security Gateway

Interfaces access control systems with other platforms such as BAS, CCTV, or visitor management systems.

### Credential Controller

Validates user credentials against an access database, managing rights, permissions, and access schedules.

## Common Audiovisual (AV) Terms and Descriptions

### AV Processor

Central controller for routing audio and video signals, managing presets, and coordinating device commands.

### DSP (Digital Signal Processor)

Handles audio mixing, microphone control, equalization (EQ), and acoustic adjustments for AV systems.

### AV Controller

Manages displays, input switching, audio levels, lighting cues, and room system presets.

### Control Processor

Dedicated automation processor, typically from platforms like Crestron or AMX, coordinating multiple AV devices and room functions.

### Room Scheduling Interface

Connects the AV system to booking platforms, allowing integration of room reservations with automated AV presets and lighting control.

## Common Fire Alarm Terms and Descriptions

### Fire Alarm Control Panel (FACP)
Central logic panel that initiates and responds to fire events, managing detection devices, notification appliances, and emergency systems.

### Annunciator Panel
Remote interface displaying fire events and system status, allowing authorized personnel to acknowledge, silence, or reset alarms.

### Control Module
Manages output logic, such as shutting down fans, releasing magnetic door holds, or activating smoke control systems.

### Input Module
Translates incoming signals from external systems or field devices (such as water flow switches or manual pull stations) into the fire alarm system.

### Network Node
Extends the FACP network to additional zones, areas, or buildings, enabling centralized monitoring and control over larger facilities.

---

## Cross-Silo and Manufacturer Branding Examples

### Johnson Controls
Common Level 2 Terminology: Field Equipment Controller (FEC), Variable Air Volume Modular Assembly (VMA)

### Siemens
Common Level 2 Terminology: PXC Controller, TX-I/O Modules

### Honeywell
Common Level 2 Terminology: Spyder Controllers, CIPer Controllers, ComfortPoint Controllers

### Crestron / AMX (Audiovisual Systems)
Common Level 2 Terminology: Control Processor, Touch Panel Interface

### Lutron (Lighting and Shades)
Common Level 2 Terminology: QSX Processor, Vive Hub, Quantum Panel

### LenelS2 (Access Control Systems)
Common Level 2 Terminology: Intelligent Controller, LNL Panel

---

### Why It Matters (More Than You Think)

**Different names cause integration breakdowns.**
Mislabeling leads to missed connections, duplicated devices, conflicting logic—and costly callbacks.

**Purdue Leveling creates a shared language.**
Using terms like "Level 2" provides a neutral way to describe where logic lives, independent of trade or brand.

**Room Controller does not mean the same thing to everyone.**
If your AV team and HVAC team both claim they are

"controlling the room," a turf war is inevitable. Integration starts with a shared understanding.

---

## Example: "Room Controller" Across Different Trades

---

### HVAC – Zone and Unit Controllers

Devices that control VAVs, fan coils, heat pumps, or radiant systems.
May include integrated sensors, scheduling, and network communications.

**Johnson Controls**
TEC Series, FEC/VMA (TEC = Thermostat with control; FEC = Configurable controllers)

**Siemens**
Room Automation Stations (PXC4/5), DXR (Room controllers with integrated I/O, sensors, and network communication)

**Honeywell**
Spyder, CIPer (Spyder is legacy; CIPer is IP-native)

**Delta Controls**
DAC/DSC Series (BACnet-native controllers within the enteliWEB system)

**Distech**
ECB/ECY Series (Supports BACnet/IP/Cloud integration with embedded scheduling)

**Trane**
UC/MP Series, Pivot Smart Thermostats (VAV-focused controllers)

**Reliable Controls**
MACH-ProZone and Space-Sensor integration (Focus on open protocols and scalability)

---

### Lighting Control – Room-Based Platforms

Devices that manage lighting scenes, switching, dimming, and occupancy/daylight input.

**Lutron**
Vive, Quantum QSX, Energi TriPak (Vive is wireless; Quantum is enterprise-grade)

**Crestron**
Zūm Room Controllers (Wireless controls combined with sensors and keypads)

**Legrand / Wattstopper**
Digital Lighting Management (Room-based logic with plug-and-play sensors)

**Acuity**
nLight AIR or Wired Room Controllers (Integration with occupancy and daylight sensors)

**Eaton**
Wavelinx Room Controller Modules (Bluetooth Low Energy, 0–10V dimming, relay modules)

**Philips / Signify**
Dynalite Room Control Units (Used in hospitality and luxury projects)

---

## Audiovisual (AV) – Experience Systems

Processors or touchpanel-based systems that execute AV scenes and room automation presets.

**Crestron**
CP4N, MPC3, or integrated TSW panels (Full AV, lighting, and shading control)

**AMX (Harman)**
NX Series Controllers (Popular in education and government applications)

**Extron**
IPCP Pro Series, MediaLink Controllers (AV control via RS-232, IR, relays, and LAN)

**QSC**
Q-SYS Core processors (Primarily audio-focused, but can control room systems)

---

## Cross-Domain – All-in-One Room Integration Platforms

Systems that simultaneously manage HVAC, lighting, AV, and shading in a unified platform.

**Crestron**

DM NVX with CP4 or TSW-X75 and AirMedia (Enterprise conference and education platforms)

**Distech**

ECLYPSE Room Control Solution (Combines HVAC, lighting, shades into one platform)

**Siemens**

DXR2 Room Controller with Desigo CC (Integrated room automation)

**Beckhoff**

Embedded PC with TwinCAT (Highly flexible controllers used for complex room environments)

---

## Why This Comparison Matters

This breakdown shows that "Room Controller" can mean very different things depending on the discipline:

- **HVAC controllers** may manage air, heating, and occupancy sensors.

- **Lighting controllers** manage brightness, daylighting, and scenes.

- **AV controllers** manage microphones, screens, and projectors.

- **Cross-domain controllers** attempt to unify multiple systems in one device—but trade-offs often exist.

When talking to a general contractor or project team, you must ask:

"Are we using the same 'Room Controller' for HVAC, AV, and lighting?"

"Is the AV 'Room Controller' going to run the HVAC?"

If no one maps **functions to devices**, confusion and conflict are guaranteed.

---

### Field Tip: Don't Just Match Labels—Map Functions

Two devices labeled "Room Controller" might have entirely different scopes.

- One might dim lights.
- The other might run a PID loop for radiant floor heating.

Always map **what each device does**, not just what it's called.

---

### Bonus: Skits and Examples

Make sure to read the skits:

- "What's in the Room?"
- "Pick a Controller, Any Controller"
- "This Isn't a Server Room"

You should be just as confused (and amused) as I was when writing them.
Be kind—I'm not a professional comedian, I just pretend to be one when explaining integration.

**What's in the Room?**

**Mac:** So you're commissioning the conference room today, right?

**Lou:** Yep. But I gotta ask... what's in the room?

**Mac:** The Room Controller handles it.

**Lou:** Okay, but what device?

**Mac:** The Room Controller.

**Lou:** Right, but what is it? Is it HVAC, AV, or lighting?

**Mac:** Yes.

**Lou:** ...Yes? So which one?

**Mac:** The Room Controller.

**Lou:** The Room Controller does all of it?

**Mac:** Depends on the manufacturer. Could be Distech, Crestron, Siemens...

**Lou:** So Distech's Room Controller controls what?

**Mac:** HVAC, lighting, shades.

**Lou:** And Crestron's?

**Mac:** Lighting, AV, scheduling.

**Lou:** Siemens?

**Mac:** Just HVAC. Unless it's integrated.

**Lou:** So when I walk in and want to turn on the lights, who do I talk to?

**Mac:** The Room Controller.

**Lou:** But which Room Controller?

**Mac:** The one that's networked.

**Lou:** And if it's not?

**Mac:** Call IT. Probably a VLAN issue.

**Lou:** Okay, fine. So the wall sensor talks to the Room Controller?

**Mac:** Not always. Sometimes it talks to the Lighting Controller.

**Lou:** I thought the Room Controller was the Lighting Controller?

**Mac:** Not necessarily. Unless it's part of the Room Pack.

**Lou:** Room Pack?

**Mac:** That's a Room Controller, Lighting Panel, and AV Gateway all in one box.

**Lou:** You're making this up.

**Mac:** I'm telling you, it's standard OT stuff.

**Lou:** So who controls the temperature?

**Mac:** The Room Controller.

**Lou:** And the lights?

**Mac:** Also the Room Controller.

**Lou:** And the projector?

**Mac:** That's the AV Processor. Unless it's tied into the Room Controller.

**Lou:** How do you people live like this?

---

## Pick a Controller, Any Controller

**Riley:** Okay, Tasha—we've got six submittals here for "Room Controllers." You want to tell me how many rooms we're building?

**Tasha** (*flipping through drawings*): Uh... twenty-four?

**Riley:** Then why do I have:

- One from HVAC for a "Zone Controller"

- One from AV labeled "Room Controller"

- One from Lighting that says "Digital Room Controller"

- One from Security that says "Access Node – Room Level"

- One from Fire Alarm called "Input Module – Room Override"

- And this one from Siemens just says "Room Automation Station – Model DXR2" like I know what that is?

**Tasha:** Oh yeah, those are all legit. Different scopes.

**Riley:** But they all say they control the same room.

**Tasha:** Well, they do. In different ways.

**Riley:** Okay. So who actually controls the room?

**Tasha:** Depends on what you mean by "controls." HVAC does temperature. Lighting does scenes. AV does presets. Access handles the doors. And fire alarm overrides everything.

**Riley:** So who's in charge?

**Tasha:** Technically, none of them. Unless we assign a master Room Controller.

**Riley:** Which one's that?

**Tasha:** Depends who wins the coordination meeting.

**Riley:** I'm not joking, Tasha. I have to approve these submittals.

**Tasha:** Then approve all of them. They'll sort it out in commissioning.

**Riley:** That's what you said about the IT racks—and we're still missing three VLANs.

**Tasha:** Well, if it helps, Crestron says their Room Controller can do HVAC too.

**Riley:** Why would I trust a touch panel with the boiler loop?

**Tasha:** Don't worry, the HVAC guy disabled that in firmware.

**Riley:** I need a drink.

**Tasha:** Add that to the punch list.

These aren't just punchlines—they're the conversations we overhear (or live through) on almost every project. The confusion around Level 2 devices isn't a technical issue—it's a language issue. That's why Purdue Model leveling matters. It gives us a shared map. And a shared map makes for better planning, fewer clashes, and fewer 11th-hour VLAN requests.

### Sidebar: What We Should Have Called It

### Device Name from the Field vs. What It Actually Is

**Thermal Brainbox**
Direct Digital Controller (DDC)

**Light-o-matic Scene Processor**
Lighting Room Controller

**Projector Preset Commander**
AV Macro Control Processor

**Conference Room Wizard**
Crestron Control System plus HVAC integration

**Portal Node Type-X**
Door Controller with Badge Reader

## Field Tip

Do not get creative with naming.
Call devices what they do—and link them clearly to Level 2 in the Purdue Model.

## Executive Summary

**Level 2 is where building systems come alive.**
It is the layer where logic runs, devices react, and real-time decisions are made across HVAC, lighting, AV, security, and more.

Yet every trade uses different terminology to describe similar devices.
One room may include five different "Room Controllers" from five vendors—all claiming ownership of the same space.
This confusion leads to duplication, integration failures, and costly delays, especially during commissioning or turnover.

## 1. Standardize on Function, Not Labels

Do not rely on trade-specific or vendor-specific names like "Room Controller" or "AV Processor."

Instead:

- Define devices by their function (local logic, I/O control, protocol translation).

- Map each device to its corresponding Purdue Model Level 2 role.

A shared functional language cuts through silos, reduces miscommunication, and aligns designers, installers, IT teams, and owners around clear integration expectations.

---

## 2. Coordinate Cross-Trade Control Logic Early

Most project delays and conflicts at Level 2 are avoidable with early coordination.

Align on:

- **Point Ownership**: Who commands which devices and points?

- **Priority Logic**: Who wins when systems send conflicting commands?

- **Device Sharing**: Are devices like $CO_2$ sensors being shared across HVAC, lighting, or security systems?

Without early cross-trade planning, your "smart building" risks becoming a battlefield—where systems fight each other and the owner pays the price in change orders and reprogramming.

# Chapter 9: The Physical Layer – Levels 1 and 0

If Level 2 is the local brain, Levels 1 and 0 are the nervous system and the muscles.

These two levels of the Purdue Model are where digital meets physical—and yet they are the most overlooked when it comes to integration strategy. While execs and IT teams live at Levels 4 and 5, and integrators thrive at Level 3, it's at Levels 1 and 0 where things get real: voltages, wires, devices, and the physical process itself. This chapter provides a quick but essential overview.

---

**Level 1: Device I/O**

This level represents the boundary between control logic (Level 2) and the physical devices that gather or execute commands.

**Examples of Inputs:**

- Temperature sensors (thermistors, RTDs, 4-20mA)
- Pressure sensors
- Occupancy detectors
- Contact closures
- Flow meters
- $CO_2$ sensors

- AV triggers (relay closures, GPIO from DSPs, IR receivers)

**Examples of Outputs:**

- Analog outputs (0-10V, 4-20mA)

- Relay contacts (on/off)

- Pulse width modulation

- Triac outputs (for SSRs or damper actuators)

- AV control triggers (projector lifts, screen relays, amplifier power control)

- Speaker-level signals and amplitude control (e.g., 70V line level, variable gain control from DSP)

**Characteristics:**

- Includes both simple I/O devices and control devices that do make decisions (e.g., terminal unit controllers), but do **not** have Ethernet connectivity. If the device performs logic but only communicates over MS/TP, LON, or other non-IP protocols, it belongs at Level 1 in this model.

- Every physical signal—whether a sensor reading or control output—ultimately connects to a controller's I/O point. If a sensor is miswired or a signal misinterpreted, it can propagate false data or commands that disrupt the entire system above.

**Integration Notes:**

- Not typically exposed to the network, but mistakes here can ripple upward.

- Standardization matters: labeling, wire colors, calibration procedures.

---

**Level 0: The Physical Process**

Level 0 is the physical environment the BAS is trying to influence. This is where commands become motion, airflow, temperature, lighting, or sound.

**Examples:**

- Air handling units, pumps, chillers, boilers

- Lighting fixtures and ballasts

- Door strikes, motors, actuators

- Valves and dampers

- Variable frequency drives (VFDs)

- AV equipment: Projectors, screens, audio DSPs, matrix switchers

- AV speakers: passive or powered—if it makes sound, it's Level 0 (unless it's a smart speaker with an Ethernet port)

**Characteristics:**

- Often has the longest lifespan in the building: 20-40 years is common.

- Receives commands from Level 1 outputs but doesn't interpret logic.

- Rarely networked (though VFDs and some fixtures can report diagnostics via BACnet/IP or Modbus).

- In AV systems, Level 0 includes devices that physically generate light, sound, or motion—not the brains that manage them.

**Integration Notes:**

- Misunderstood or undocumented Level 0 devices can render higher-level systems useless.

- Be sure of voltage requirements, control signal types, and failure modes (open vs. closed, power fail behavior).

- For AV, ensure compatibility between control signal types (e.g., dry contact relay vs. voltage-driven trigger).

---

## Why These Levels Matter

Too many integration strategies start at Level 3 or above and treat Levels 1 and 0 as someone else's problem. But no amount of API magic can fix a miswired thermistor or a projector lift that only works when the screen is already down.

**The Risks:**

- Commissioning delays due to sensor calibration errors

- Energy waste from stuck valves or relays

- Safety hazards from misfired outputs or incorrect wiring

- AV failures where display power, input switching, or audio routing depends on Level 1 contact closures being in sync

**The Best Practices:**

- Label everything. Seriously. And use consistent, job-wide standards.

- Provide pluggable terminal blocks when possible to ease service and upgrades.

- Validate input types during startup (e.g., 10k type 3 thermistor vs. 20k).

- Include Level 1/0 checks in integration test plans.

- For AV: Coordinate wiring plans with control logic—especially when relying on serial, IR, or GPIO.

---

## Quick Summary – Levels 1 and 0

Levels 1 and 0 may not appear on executive dashboards or cloud platforms, but they are the foundation where every sensor signal, actuator movement, and AV trigger begins.

- **Level 1 – Device I/O**:
  Sensors, relays, analog and digital signals, and AV triggers.

- **Level 0 – Physical Process**:
  HVAC units, lighting fixtures, actuators, variable frequency drives (VFDs), and AV output devices such as speakers.

---

These levels are the physical realities beneath strategy— and they matter deeply.

The work at these levels must be performed by skilled professionals such as AV designers, RCDDs, HVAC controls engineers, lighting specialists, and commissioning agents. However, their efforts must align with the **integration strategy and architecture** defined at the higher levels.

A smart building does not start with a dashboard.
It starts with a solid physical foundation that honors and supports the strategy above it.

---

### Executive Summary – Quick Recap

**Levels 1 and 0 form the foundation of control systems.**
They are where digital logic meets the real world—through sensors, relays, actuators, and operational equipment—yet they are often overlooked during integration planning.

**Mistakes ripple upward.**
A miswired sensor, mislabeled relay, or incompatible AV

trigger can cause delays, inefficiencies, or failures that no high-level software platform can fix after the fact.

**Integration starts at the wire.**

Labeling, validating signal types, confirming physical compatibility, and coordinating wiring plans must be part of the initial integration strategy—not just left to execution in the field.

# Chapter 10: HVAC Control Systems

HVAC systems are often the largest and most mature domain within the world of building controls. But maturity doesn't always mean integration-friendly. In many projects. HVAC is often the first system on a project to bring controls to the table—leading many buildings to revolve around HVAC-centric logic, while leaving other systems like AV and security struggling to catch up or integrate after the fact. Lighting and fire alarms integrate well with HVAC; AV and security, not so much. Some of the specialty systems don't even know that HVAC exists, and vice versa.

This chapter is about understanding HVAC controls within the context of an integrated building strategy—how they work, what they control, and how they need to play nicely with others. While it is the big brother of the controls family there are a lot of cases where it should not be in the prime position, but there are many times when it really should be anyway.

---

## HVAC Control Fundamentals

At its core, HVAC control is about maintaining environmental conditions—temperature, humidity, airflow, and sometimes pressure—within set parameters. Control systems typically:

- Read sensor inputs (temperature, humidity, $CO_2$, etc.)

- Execute logic or PID loops in DDC controllers (Level 2)

- Send commands to outputs (dampers, valves, fans)

- Report values and alarms upstream to the BAS or integration platform (Level 3)

Most HVAC logic is based on proportional-integral-derivative (PID) control, which continuously adjusts outputs (like dampers or valves) to maintain a setpoint based on sensor feedback.

System types controlled:

- Air Handling Units (AHUs)

- Variable Air Volume (VAV) boxes

- Chillers and boilers

- Fan coil units

- Heat pumps and VRF/VRV systems

---

## Common HVAC Components in Control Systems

### Temperature Sensor
Provides input for space or duct temperature feedback, enabling control sequences to maintain desired conditions.

### Actuator
Drives the mechanical movement of dampers or valves based on control logic outputs.

### Controller (DDC)
Runs logic sequences, manages device communication, and controls field devices within the HVAC system.

### Thermostat
Acts as a user interface at the zone level, allowing occupants or facility teams to set or adjust temperature and mode inputs.

### Variable Frequency Drive (VFD)
Modulates motor speed for fans or pumps to improve energy efficiency and system responsiveness.

### BACnet MSTP/IP
Common open communication protocol enabling controllers, sensors, and devices to share data across HVAC networks.

---

### Points of Integration

HVAC doesn't operate in a vacuum. Good design considers how HVAC systems respond to, or influence, other building systems:

- **Lighting**: Shared occupancy sensors or scheduling coordination

- **Fire Alarm**: Smoke control, AHU shutdowns, damper actuation

- **Security**: Door position triggers temperature setbacks

- **Shades**: Reduce solar gain to improve cooling efficiency

- **AV**: Coordinate pre-cooling or shutdown with event schedules

- ➤ Note: *Integration logic must consider system priorities—life safety systems like fire alarm always take precedence over comfort logic.*

When possible, HVAC points should be mapped for visibility in the shared Level 3 system, using clear naming conventions and normalized units.

---

## Naming Conventions and Point Lists

A solid point naming strategy improves commissioning, integration, and analytics. Consider:

- **System.Zone.PointType** (e.g., AHU1.Zone5.SupplyAirTemp)

- Standard abbreviations (SAT, RAT, SP, OA%)

- Units and scaling (C/F, %, CFM, kW)

Include a point list early in design. Make sure the integrator knows what's being exposed—and why. Within your cybersecurity rules, determine what it readable, writable, or there can be another state of requested and an operator has to approve.

---

## Graphics and User Interface Expectations

HVAC is often the most visible system in the BAS. Poor graphics or unreadable alarms can ruin operator trust. Best practices:

- Show floorplans and system diagrams

- Animate airflows or temperature zones

- Color-code alarms and overrides

- Include drill-downs for equipment

Remember: if the graphics are ugly or hard to navigate, no one will trust the system. Good UI builds operator confidence. AV is usually the king of good UIs on an integrated project. Try to take advantage of this where you, and certainly don't let one silos graphics fall too far from the ideal for the project.

💡 **Tip:** HVAC should be designed for **collaboration**, not **dominance**. Align zone logic with room usage, not just ducts and sensors. If AV or lighting owns the user interface in a room, HVAC should follow—not fight it.

---

## HVAC Trends in Modern Integration

Modern HVAC control design must also contend with:

- **Cloud-based analytics platforms** for fault detection

- **Smart thermostats** with mobile access

- **Demand response integration** for energy curtailment

- **Wireless sensors** in retrofit or flexible space environments

- **Decarbonization goals** driving electrification and sequencing changes

---

## Coordination Tips

- HVAC control scope should be coordinated with mechanical scope of work—ensure clear handoff between installer and integrator

- Confirm sensor placement with architect and lighting designer (avoid redundancy)

- Align startup and test schedules with other trades

- Document control sequences in Division 23 and reflect in Division 25

- For multi-building campuses or packaged HVAC systems with their own controls, coordinate integration paths and ownership models early.

HVAC often sets the tone for a project. Design it for integration—not just comfort—and it becomes the backbone of a smart building, not just its lungs.

In the next chapter, we'll move from climate to current—**lighting control systems**, and how they shape both energy usage and occupant experience.

## Executive Summary

**HVAC is the backbone of most building control systems.**
It is often the first system installed and the most mature—
but its dominance can lead to unbalanced integration if
other systems are treated as afterthoughts.

**Effective integration requires HVAC to share data and
coordinate with other systems.**
Lighting, fire alarm, security, shading, and audiovisual
platforms must connect through clear naming conventions,
mapped points, and a unified Level 3 interface to avoid silos
and miscommunication.

**Modern HVAC control strategies must balance multiple
demands.**
Today's systems must support comfort, energy efficiency,
mobile access, fault detection, demand response, and
decarbonization goals.
Successful integration requires these evolving capabilities
to align not only within HVAC but across all building
systems under a unified integration strategy.

# Chapter 11: Lighting Control Systems

Lighting control is one of the most widely deployed—and inconsistently integrated—systems in modern buildings, often treated as separate from the BAS even though it touches nearly every occupied space. It sits at the intersection of energy savings, occupant comfort, aesthetics, code compliance, and automation.

This chapter examines how lighting control systems operate, where they should integrate, and what makes them succeed or fail within the broader building controls strategy.

## Fundamentals of Lighting Control

Lighting systems have shifted dramatically in the last 20 years—from relay-based panels and analog dimming to digital, addressable, and often wireless platforms. Today's systems include:

- Scheduling (based on time, occupancy, or daylight)
- Dimming and scene control
- Astronomical clocks and daylight harvesting
- Occupancy/vacancy sensing
- Plug load control
- Integration with shading, AV, or security

## Lighting Control Systems – Overview

Lighting controls can function as standalone systems—or be deeply integrated into BAS and Level 3 supervisory systems.

## Typical Lighting Control Components

### Lighting Control Panel
Houses relays or dimmers and may run basic or advanced lighting control logic for multiple zones.

### Fixture Controller
Manages individual light fixtures, allowing addressable control for dimming, color tuning, or scheduling.

### Room Controller
Local controller typically installed in a room or zone; integrates occupancy sensors and scene control logic.

### Occupancy Sensor
Detects presence or absence in a space and triggers lighting zones accordingly for energy savings and user comfort.

### Photocell
Measures daylight levels to enable automatic dimming or shading adjustments based on available natural light.

### Wall Station or Switch
Manual control point for users to trigger lighting scenes, adjust brightness, or override automated settings.

## Code Drivers and Industry Standards

Lighting controls are frequently dictated by code. Key references include:

- **ASHRAE 90.1** and **IECC** (International Energy Conservation Code)

- **Title 24** (California)

- **UL 924** for emergency lighting

- **DALI**, **0–10V**, and **DMX** protocols for control

Design must balance compliance, usability, and integration opportunities.

## Integration Opportunities

Lighting is a natural integration candidate because it affects every space. Common strategies include:

- **Shared sensors** for HVAC and lighting (occupancy)

- Unified **scene control** not only improves occupant experience but also reduces interface clutter across touchpanels or user apps.

- **After-hours HVAC based on lighting state**

- **Fire alarm events** turning on egress lighting

- **Access control triggers** lighting in response to badge reads

Lighting should be mapped into the Level 3 platform with intuitive grouping and labeling.

## Lighting Control Topologies

There's no single standard layout. Common approaches:

- **Centralized panel**: All wiring goes back to a single panel

- **Room-based systems**: Each room has a controller, often wireless

- **Fixture-integrated**: Logic lives inside each luminaire

- **Hybrid**: Combines centralized and distributed

Choose your topology based on building type, wiring constraints, retrofit limitations, and how the system will interface with IT or networked infrastructure.

## Integration Challenges

- **Multiple vendors with misaligned deliverables**— fixture suppliers may specify drivers that don't match the intended control protocol

- **Complex commissioning** due to local and global logic

- **DMX and DALI incompatibilities** without proper gateways

- **Manual overrides conflicting with scheduled logic**

- **Graphical interfaces** that lack operator-friendly grouping

Successful lighting integration comes down to three things: clarity in design, consistency in execution, and clearly defined ownership from day one.

**Best Practices**

- Use unique, clear naming for lighting zones and devices

- Coordinate with AV, shading, and HVAC early in design

- Document the lighting sequences in Division 26 or 25

- Require a full point list of what is visible in the BAS

- Validate control zones on-site during construction mock-ups

- Ensure the lighting control specification is bundled with the fixture package to avoid costly mismatches during procurement.

When done right, lighting controls contribute not just to energy performance—but to occupant wellbeing and intuitive building operation.

### Procurement Pitfall – Why Lighting Control Specs Must Travel with the Fixtures

### Real-World Insight:
In my experience owning a lighting controls firm with my partner Bill—which we later sold to Crestron Electronics, where it became their service department—we saw

firsthand how tightly lighting controls are tied to the fixture package.

In most construction projects, lighting controls are not treated as a standalone system.
They are packaged with the fixtures, selected by the lighting representative, and purchased through the same distribution channels.
This bundling is deeply embedded in how rep firms and manufacturers operate—and nearly impossible to separate once the fixture package is finalized.

---

**Field Example:**
On one project, the lighting controls specification was not properly tied to the fixture package. Two days before turnover, we were called in to make the lighting controls work. The problem: the newly selected fixture drivers did not support the control protocol shown in the original drawings.

We were able to get the system functional, but only with **significantly reduced capability**. A beautiful, high-end restaurant ended up with lights that dimmed in clunky, uneven steps—because the fixture controls were mismatched to the specified system. The designer's vision was compromised simply because controls were not locked in with the fixtures early enough.

---

**Lessons Learned**

- **Coordinate early** between the lighting designer, controls integrator, and electrical engineer.

- **Include controls requirements** directly in the fixture schedule.

- **Avoid vague "by others" placeholders** in Division 26 without clarifying responsibilities.

- **Treat controls as infrastructure**, not as accessories or afterthoughts.

---

Lighting controls cannot be tacked on after fixture selections are made.
If you wait, you are designing controls in theory—and hoping for compatibility later.
Smart projects specify controls and fixtures together from the beginning.

In the next chapter, we'll shift focus to systems that are less frequently integrated—but just as critical: **fire alarm and life safety**.

**Executive Summary**

**Lighting controls are everywhere—but often siloed.**
Although they touch every space in a building, lighting control systems are often isolated. Yet they bridge energy efficiency, aesthetics, occupant comfort, and code compliance, making them prime candidates for deeper integration.

**Modern lighting systems are highly capable—but require careful coordination.**

Functions like dimming, daylight harvesting, scene control, and occupancy-based automation are now standard features.

Successful integration demands clear coordination across vendors, communication protocols, and design disciplines.

**Effective integration hinges on early planning and structure.**

Lighting integration requires early coordination, consistent zone naming, and appropriate topology selection.

These steps ensure that lighting operates seamlessly with HVAC, AV, shading, and security systems through a shared Level 3 supervisory platform.

**Lighting control strategies must travel with the fixture package.**

Early coordination between lighting designers, electrical engineers, and controls integrators ensures protocol compatibility and preserves the intended functionality and occupant experience.

# Chapter 12: Fire Alarm and Life Safety Integration

Fire alarm systems are often considered "untouchable"—separate from other systems, tightly regulated, and not to be interfered with. And for good reason: they are life safety systems with legal, code-driven requirements and very little margin for experimentation. But that doesn't mean they can't be integrated. In fact, strategic, code-compliant integration of fire alarm systems is essential to a fully coordinated building.

This chapter explores how fire alarm systems interface with other building systems, what is permitted (and not permitted) by code, and how to design integrations that enhance safety without adding risk.

---

## Understanding the Role of the Fire Alarm System

The Fire Alarm Control Panel (FACP) acts as the central hub for all fire detection and response logic in a building. It connects to:

- Smoke detectors, heat detectors, manual pull stations

- Notification devices (strobes, speakers, horns)

- Elevator recall systems

- Smoke control systems

- Fire suppression interfaces (e.g., pre-action systems)

In many facilities, the FACP is also connected to the BAS or other systems—but **only under strict rules.**

---

### Common Integration Points – Permitted by Code

When designed correctly, integration between fire alarm systems and other building systems is both common and allowed, typically governed by NFPA 72 standards.

Typical integration points include:

### HVAC Systems
Trigger air handling unit (AHU) shutdowns, damper closures, and post-event fan purge sequences to contain or expel smoke.

### Access Control Systems
Unlock magnetic locks and disable door strikes during an alarm to allow safe egress.

### Lighting Systems
Turn on emergency egress lighting when an alarm condition is detected, ensuring pathways are visible for occupants.

### Elevator Systems
Initiate elevator recall to designated safe floors to prevent occupants from using elevators during a fire.

### Building Automation System (BAS) Graphics
Display alarm statuses, event logs, and system conditions on operator dashboards to assist emergency response and building management.

**Mass Notification Systems**

Trigger public address announcements, strobes, or other emergency alerts to occupants as part of evacuation or shelter-in-place procedures.

These integrations are typically **one-way** from fire alarm to the other system, and they must function **even if the BAS is down.**

While not required it is highly recommended that audiovisual system be integrated as well to duck or mute the audio from the fire alarm when needed.

---

**Integration Best Practices**

- Use **dry contact relays or supervised inputs** between FACP and other systems

- Keep integrations **read-only** into the FACP—no system should write commands to the fire alarm

- Maintain **separate pathways** for fire alarm wiring to meet code

- Ensure all programming and relay actions are documented in the **Sequence of Operations** and reviewed by the **AHJ (Authority Having Jurisdiction)**

---

**Common Misconceptions**

- "The BAS can shut off fire alarm speakers." → **False.** Only the FACP or fire department can control notification circuits.

- "I can read status from the FACP over BACnet." → **Only if implemented through an approved gateway and read-only structure.**

- "We can share sensors." → **Sometimes, but rarely advisable.** Always verify with local code and AHJ.

---

## Who's Responsible?

Responsibility is often split:

- **Fire Alarm Contractor** provides relay contacts or integration modules

- **BAS Contractor** wires to those contacts and triggers sequences in their system

- **Electrical Contractor** may install interface wiring and ensure correct conduit

- **Commissioning Agent** verifies correct operation during integrated tests

Clarify responsibilities in **Division 25**, and coordinate closely during shop drawing and commissioning phases.

---

## Design Documentation

Good documentation avoids confusion and failed inspections:

- Drawings should show relay wiring and control logic clearly

- Sequences should list both initiating device and expected downstream response

- FACP integration diagrams should be reviewed by the design engineer and AHJ

Use standardized naming like: `FACP_AlarmToAHU1_Shutdown`

---

**Real-World Insight – Fire Alarm Was My First Introduction to Controls**

One of my first jobs in the controls industry was working as a fire alarm technician.
I performed monthly maintenance checks, responded to alarms across campus, and coordinated with the vendor to update the fire alarm system.
It was my first real exposure to just how much **documentation, discipline, and detail** fire alarm systems require.

Later, we attempted a one-way integration from our Simplex fire alarm system into a Siemens HVAC BAS.
The idea was simple: allow the BAS to **monitor** alarm and trouble signals—without ever sending control commands back to the fire alarm.

It was strictly read-only:

- The fire alarm could send data outward.

- No signals were allowed to return into the fire alarm panel.

That experience stayed with me.
It was an early and lasting lesson in **why boundaries matter** in system integration—especially when **life safety systems** are involved.

### Summary

Fire alarm systems are foundational to life safety, but they don't have to operate in isolation. With proper planning, code-aware implementation, and clearly defined integration points, fire alarm systems can contribute to a coordinated, safer, smarter building.

In the next chapter, we'll look at another critical safety layer: **Security and Access Control**—and how those systems interact with the rest of the building stack.

### Executive Summary

**Fire alarm systems are critical to life safety—and highly regulated.**
This sometimes makes them seem "off-limits" for integration. However, with careful design and strict code compliance, fire alarm systems can and should coordinate with other building systems to improve occupant safety and operational response.

**Permitted integration points are well-defined.**
Connections to HVAC shutdowns, egress lighting activation, access control overrides, elevator recalls, and BAS graphics are common.
These integrations typically use **one-way, read-only** communications from the Fire Alarm Control Panel (FACP) to other systems, ensuring that fire alarm functions are not compromised.

**Successful fire alarm integration requires clear boundaries and thorough documentation.**
Integration must use dry contacts or supervised inputs, strictly avoid writing commands to the FACP, coordinate responsibilities across trades, and ensure all integration points are reviewed and approved by the Authority Having Jurisdiction (AHJ).
The full sequence of operations should clearly document every integration interaction.

# Chapter 13: Access Control Systems

Access control is not just about keeping unauthorized people out—it is about managing who can go where, when, and under what conditions.

It is the digital equivalent of keys and doormen, and when designed well, it becomes the backbone of both security and building automation.

In a smart building, access control does not live in isolation. It coordinates with HVAC, lighting, audiovisual (AV), and life safety systems to enhance occupant experience and operational efficiency.

---

## Core Components of Access Control

### Access Control Panel
Acts as the logic engine for credential validation and door control.

### Card Reader
Reads badges or mobile credentials to authenticate users.

### Magnetic Lock or Electric Strike
Provides the physical mechanism to lock or unlock doors based on credential validation.

### Door Contact
Monitors door status to confirm whether it is open or closed.

**Request-to-Exit (REX) Device**
Detects occupant presence to allow safe egress without triggering alarms.

---

## Credential Types

- **Physical Credentials**: Proximity cards, smart cards, and key fobs.

- **PIN Codes**: Keypad-based codes, often combined with cards for multi-factor authentication.

- **Biometric Credentials**: Fingerprint, iris scan, and facial recognition, commonly used in high-security or high-traffic areas.

- **Mobile Credentials**: Bluetooth Low Energy (BLE) and Near-Field Communication (NFC)-based mobile app credentials.

- **Dual Credentialing**: Combining two factors (such as card plus PIN) for enhanced security.

---

## Access Logic Design

- **Time Zones and Schedules**: Define when a credential is valid.

- **Anti-Passback**: Prevents re-entry into a secure area without exiting first, maintaining access integrity.

- **Lockdown Modes**: Enable one-button overrides to secure areas during emergencies.

- **Security Zones**: Logical groupings of access-controlled spaces for easier management.

- **Credential Hierarchies**: Differentiate permissions for visitors, employees, and executives.

---

## Integration Opportunities

Smart access control systems can trigger coordinated actions across the building:

### Badge Scan at Door

- Turns on lights and sets the HVAC zone to occupied mode.

- Helps reduce energy waste by only conditioning spaces when in use.

### Executive Badge at Elevator

- Triggers AV system startup in the boardroom or designated executive area.

- Enhances occupant experience by streamlining high-priority workflows.

### Door Forced Open

- Sends an alarm to the Building Automation System (BAS).

- Activates lighting strobes or AV messaging to alert security staff.

### Fire Alarm Event

- Releases all magnetic locks.

- Required by NFPA and IBC codes to ensure safe occupant egress during emergencies.

---

## Naming Conventions Matter

### Bad Examples

- Reader1

- MagLockA

- ZoneX

### Better Examples

- DOOR_103_ReaderEntry

- DOOR_103_MagLock

- LEVEL_2_EAST_Corridor_Access

Clear, consistent naming helps integration mapping, troubleshooting, and training.

---

## Commissioning Checklist for Access Control

### System Configuration

- Confirm credential database is current and loaded.

- Test time zones and holiday schedules.

**Hardware and Functionality**

- Verify Request-to-Exit (REX) devices and door contacts operate correctly.

- Validate fail-safe and unlock behavior under emergency conditions.

**System Integration**

- Cross-check naming conventions across integrated systems.

- Test alarm outputs such as BAS alerts, lighting strobes, and AV messaging.

---

**Integration Pitfalls to Avoid**

- **Uncoordinated Time Zones**: Systems behaving inconsistently, such as doors unlocking when they should not.

- **Inconsistent Door Naming**: Integration maps break or require rework.

- **No Priority Logic**: Two systems attempt to control the same door simultaneously.

- **Overwriting AV or HVAC Triggers**: Without proper failback logic, system behavior becomes unreliable.

- **Lack of Early Coordination**: Integration must be planned alongside HVAC, lighting, and security—not as an afterthought.

(See Appendix I for more on how Division 28 systems interface with Division 25 integration requirements.)

---

**Real-World Insight – Schedules Need People in the Loop**

I once worked on a project where the HVAC system had a holiday schedule correctly set to "Off"—but the access control system still had door schedules active for an event center.
That morning, the building was technically closed, but the doors automatically unlocked.
HVAC zones fired up unnecessarily, and the facility looked open for business when it was not.

The result?
Wasted energy, security vulnerabilities, and a lot of operator confusion.

The lesson:
Even the smartest buildings need human judgment.
Schedules must be synchronized across systems—but certain actions, like unlocking doors, should always require human confirmation.
A facility should not automatically open itself just because the calendar says it might.

---

## Executive Summary

**Access control is more than security—it manages movement, timing, and permissions across a building.**
Access control plays a critical role in smart building coordination with HVAC, AV, lighting, and life safety systems.

**Modern access systems support sophisticated logic.**
Features such as schedules, lockdowns, credential hierarchies, and anti-passback create opportunities for automation—especially when integrated into Level 3 supervisory platforms.

**Successful integration depends on structure and planning.**
Clear device naming, aligned time zones, and priority logic ensure that access control systems enhance safety, improve energy efficiency, and deliver seamless occupant experiences without introducing system conflicts.

# Chapter 14: Intrusion Detection Systems

While access control manages who should enter, intrusion detection is all about identifying who should not. Intrusion Detection Systems (IDS) are common in offices, data centers, laboratories, and mission-critical facilities— but they are often treated as an afterthought in integration planning.

A well-integrated IDS not only improves security but also enables automation across lighting, HVAC, AV, and emergency systems.

---

## Core IDS Components

### Door or Window Contact
Detects unauthorized opening of doors or windows.

### Motion Detector (PIR/Microwave)
Senses movement within protected zones or corridors.

### Glass-Break Sensor
Detects high-frequency shatter events indicating broken windows or glass walls.

### Panic or Duress Button
Provides a manual trigger to signal an immediate alarm condition.

### Alarm Panel
Processes IDS events locally, managing device inputs and alarm outputs.

**Keypad or Arming Device**

Allows users to arm or disarm zones and manage system states.

---

**Common Use Cases in Buildings**

- **Office After-Hours Monitoring**
  Detect motion in normally unoccupied office spaces to trigger alarms or HVAC adjustments.

- **Data Center Protection**
  Alert security if a server room door is opened without a corresponding badge scan.

- **Warehouse Perimeter Monitoring**
  Track unauthorized access through loading docks, roof hatches, or fence lines.

- **High-Security Zones**
  Combine IDS sensors with badge access, video surveillance, and audio detection for layered security.

---

**Integration with Other Systems**

**Intrusion Detected**

- HVAC setback to unoccupied mode to conserve energy and flag space as insecure.

**Motion Detected in Secure Area**

- Alert security staff, flash area lights, and activate security cameras.

**Panic Button Pressed**

- Duck AV system audio, broadcast emergency alerts through the fire alarm or public address system.

**Armed Zone Breached**

- Force door unlock and trigger building-wide alerts via the BAS.

---

**Handling False Alarms**

False alarms degrade system trust and operator response. Common causes include:

- HVAC airflow triggering motion sensors.

- Pets or cleaning staff entering armed zones.

- Misaligned or damaged door contacts.

**Mitigation Strategies:**

- Use dual-technology motion detectors (combining passive infrared and microwave detection).

- Implement time delays or verification zones.

- Integrate with access control systems to confirm motion events against badge activity.

---

## Common IDS Architectures

### Hardwired Zones

Each sensor or contact is individually wired to the panel. Best suited for high-security or permanent installations.

### Bus-Based Systems (RS-485 Architecture)

Devices daisy-chained on a shared communication wire. Efficient for medium to large commercial installations.

### Wireless Sensors

Battery-powered devices using radio communication. Ideal for retrofits, temporary spaces, or small-scale projects.

---

## Sample Integration Sequence

### Scenario:

Unauthorized entry to a server room after hours.

### Trigger:

Door contact opened without a corresponding badge scan.

### Actions Taken:

- Turn on area lighting.

- Activate and bookmark footage from the nearest security camera.

- Notify the on-call technician and the security team.

- Adjust HVAC to a neutral mode to avoid masking smoke or overheating indicators.

## Commissioning Checklist for IDS

- Confirm all door contacts are aligned and functional.

- Test motion detection zones for proper sensitivity and aiming.

- Verify panic and duress devices generate the correct event logs.

- Test arming and disarming zones from user interfaces.

- Confirm that alarms trigger the intended lighting, BAS, and video management system responses.

- Set up a mechanism for reviewing and categorizing false alarms.

## Common Integration Pitfalls

- IDS-triggered HVAC overrides conflicting with normal building schedules.

- Lack of redundancy, causing alarm events to remain trapped inside the IDS panel without wider notification.

- Poorly secured remote connections to IDS panels.

- Multiple contractors installing AV, BAS, and security systems independently without coordination, resulting in integration failures.

Intrusion detection becomes far more powerful when it is not isolated, but instead informs and collaborates with the broader building systems.

When every system can "listen" and "respond" to a threat intelligently, both safety and operational resilience improve dramatically.

## Executive Summary

**Intrusion Detection Systems (IDS) are essential for identifying unauthorized access and threats.**
They are particularly critical in sensitive areas such as offices, data centers, and mission-critical facilities. When properly integrated, IDS systems extend beyond basic security—they trigger automation responses across HVAC, lighting, and AV systems to enhance building intelligence.

**Key integration opportunities include:**

- Automating HVAC setbacks in unoccupied areas.

- Triggering multi-system alerts during a breach.

- Enabling coordinated panic responses through audio-visual cues.

**Successful IDS integration requires careful planning.**
Mitigating false alarms, ensuring alarm output redundancy, and coordinating among security, HVAC, AV, and facility

teams are essential to maintain reliability and build user trust.

# Chapter 15: Video Surveillance Systems (VSS)

Video surveillance has evolved far beyond simple crime deterrence.

Today, it serves as a forensic tool, a behavior analytics engine, and a critical sensor platform within integrated Building Automation Systems.

In a smart building, VSS acts as a shared sensor system that informs access control, intrusion detection, audiovisual (AV) systems, lighting, and even HVAC.

However, integration is often limited by poor planning, underestimating bandwidth and storage requirements, or falling into proprietary vendor lock-in.

This chapter focuses on how to make VSS a true team player in smart building integration.

---

## Core Surveillance System Components

### Camera

Captures video footage and sometimes audio. Modern IP cameras often include basic analytics. Cameras are considered Level 1 devices in the Purdue Model.

### Video Management System (VMS)

Centralizes camera feeds, handles recording, playback, bookmarking, and analytics integration. Acts as the operational brain of the VSS at Level 3.

**NVR or DVR (Network or Digital Video Recorder)**
Stores video locally, often part of the Level 3 system stack.

**Cloud Video Surveillance Systems (VSS)**
Streams video to cloud-hosted platforms, enabling remote viewing and storage without on-site servers.

**Switch or PoE Injector**
Provides power and network connectivity for IP cameras through Power-over-Ethernet (PoE).

**Encoder or Decoder**
Converts analog camera signals to digital for use in IP-based systems, useful for retrofit applications.

---

## Camera Types and Selection

**Fixed Dome Camera**
Ideal for hallways, small rooms, and unobtrusive surveillance.

**Pan-Tilt-Zoom (PTZ) Camera**
Used in parking lots, lobbies, and areas requiring remote-controlled patrols.

**Panoramic or Fisheye Camera**
Captures wide open areas with a single mounting point.

**Thermal or Infrared (IR) Camera**
Essential for monitoring in zero-light or perimeter security applications.

**Multi-Sensor Camera**

Offers multiple wide-angle views from a single mount, improving coverage efficiency.

**Design Tip:** Choose field of view and frame rate based on the event type, not just coverage goals.
Higher frame rates are not always better for low-activity zones.

---

## Integration Opportunities

**Badge Swipe at Secure Door**

Trigger a VMS bookmark capturing 15 seconds before and after the swipe to aid access investigations.

**Intrusion Detection Event**

Initiate continuous recording and alert operations staff for real-time response to after-hours movement.

**Fire Alarm Activation**

Automatically record and archive footage of building egress paths to support post-incident reviews.

**AV System Event**

Move PTZ cameras to focus on a stage or podium for large presentations or conferencing events.

**Integration Tip:** Send activity alerts to BAS after-hours based on VSS motion analytics, allowing systems to adjust lighting and HVAC settings dynamically.

---

## Storage Considerations

### On-Premises NVR/DVR Systems

- Offers local control and high performance without recurring costs.

- Requires planning for sufficient power, cooling, and physical security.

### Cloud-Based VSS Platforms

- Suitable for distributed sites with minimal local infrastructure.

- Requires careful management of bandwidth and ongoing subscription costs.

### Hybrid Architectures

- Combines local recording for quick retrieval with cloud archiving for redundancy and search capabilities.

- Increasingly popular for larger campuses.

---

### Storage Sizing Rule-of-Thumb

At 1080p resolution and 15 frames per second:

- One camera typically consumes between 1.5 to 2.5 Mbps.

- Monthly storage per camera is approximately 600 GB.

Multiply storage needs by the number of cameras, desired retention time, and RAID overhead.
Use vendor-provided calculators for precise estimates.

## Cybersecurity for Video Surveillance Systems

Treat VSS cybersecurity with the same seriousness as financial or medical systems:

- Segment VSS traffic onto dedicated VLANs.

- Prohibit direct internet access from cameras.

- Require HTTPS, RTSP authentication, and access logging.

- Disable default passwords immediately upon installation.

- Store audit logs for incident review and chain of custody validation.

A camera system accessible via a default password is not secure—it is a liability.

## Network Integration Design

### Multicast vs. Unicast
Multicast reduces switch loading but requires more configuration expertise.

### PoE Budgeting

Always verify switch power budgets, especially when deploying high-wattage PTZ or IR cameras.

### VLAN Tagging

Separate VSS traffic from BAS, access control, and other critical systems.

### Quality of Service (QoS)

Prioritize video packets to avoid dropped frames during periods of network congestion.

---

### Privacy and Access Control

- Apply strict role-based access permissions to restrict who can view sensitive camera feeds.

- Use masking or redaction techniques in areas with privacy concerns.

- Define video retention periods based on legal compliance and company policy.

- Establish chain of custody practices for evidence handling.

---

### Commissioning Checklist for VSS

- Confirm all cameras are online and correctly configured.

- Validate fields of view match design intent.

- Ensure VMS time synchronization through NTP servers.

- Test bookmarking rules, motion analytics, and alert triggers.

- Confirm cybersecurity settings and remote viewing permissions.

- Validate storage system health, including RAID and failover configurations.

- Verify time-synced integration with BAS, access control, and AV systems.

---

**Integration Summary**

**Access Control Systems**

- Bookmark entry events and trigger alerts for tailgating incidents.

**Fire Alarm Systems**

- Capture and archive evacuation footage to support post-incident analysis.

**Building Automation Systems (BAS)**

- Optionally alert on unauthorized occupancy after-hours.

**Audiovisual Systems (AV)**

- Shift PTZ presets for event coverage or meetings.

### VSS and the Purdue Model

- **Level 1:** Cameras as field devices capturing raw data.

- **Level 3:** VMS systems aggregating, managing, and exposing video feeds and analytics.

- **Levels 4–5:** Optional enterprise-level dashboards or incident management platforms consuming VSS data.

Surveillance is not just visual—it becomes a rich layer of operational awareness when fully integrated into the smart building ecosystem.

### Real-World Planning Insight

On one project, a specification called for 80 high-resolution IP cameras—but no one accounted for PoE switch loads or rack cooling.
The result: overheated network racks and frozen video streams.
Lesson learned: VSS is an infrastructure load.
Design for **power**, **heat**, and **bandwidth** the same way you would for core IT systems.

### Real-World Insight – Surveillance Without Specs = Storage Nightmare

A public agency once requested full HD surveillance, long-term video retention, and live tracking across an entire campus—but without consulting a VMS designer first. When the true storage requirement was calculated, it reached petabyte scale, requiring a full datacenter buildout just for video.

The lesson:

Video surveillance demands realistic storage, bandwidth, and infrastructure planning. Otherwise, the system collapses under its own weight.

---

## Final Thought

Surveillance is no longer just for security—it is a powerful sensor platform.

When integrated into the building's supervisory and automation systems, it delivers real-time situational awareness that supports safety, operations, and occupant experience.

But it only works if designed with purpose, integrated intelligently, and secured rigorously.

---

## Executive Summary

**Video Surveillance Systems (VSS) are vital operational sensors in modern smart buildings.**

Beyond crime deterrence, VSS platforms now feed data

into access control, intrusion detection, AV, lighting, HVAC, and emergency response systems.

**Integration opportunities include:**

- Bookmarking access events.

- Coordinating PTZ camera movement during AV events.

- Supporting post-incident reviews through archived footage.

**Effective VSS integration requires thoughtful infrastructure planning.**
Proper attention to network traffic, storage capacity, cybersecurity, and privacy ensures that video data remains available, reliable, and secure across the entire building ecosystem.

# Chapter 16: Shade, Blind, and Envelope Systems

Window coverings may seem like a minor design detail—but when integrated effectively, they become an essential component of building performance, occupant comfort, and smart building automation.

Automated shades and blinds reduce glare, control solar heat gain, protect occupant privacy, and contribute to energy efficiency—all while enhancing the occupant experience.

This chapter focuses on how automated shading systems integrate with lighting, HVAC, AV, and façade systems to create a cohesive building environment.

---

## Core Shading System Components

### Motorized Shade
Raises or lowers window treatments via low-voltage motors, either manually or automatically.

### Shade Controller
Receives commands from audiovisual, building automation, or lighting systems to coordinate shade movement.

### Wall Switch or Keypad
Provides manual override capabilities for local user control.

### Daylight Sensor

Measures available natural light and triggers automated shade adjustments.

### Scheduler

Automates shade behavior based on time-of-day, sunrise, or sunset patterns.

---

### Integration Use Cases

### Lighting Systems

Lower shades automatically when indoor daylight exceeds preset lux levels, reducing glare and balancing lighting loads.

### Audiovisual (AV) Systems

Lower blackout shades to darken rooms for projection or presentation modes.

### HVAC Systems

Lower shades to reduce solar heat gain and assist HVAC systems in maintaining comfort during peak sun exposure.

### Occupancy Sensors

Open shades during occupied hours to promote natural lighting; close shades when spaces are unoccupied to enhance privacy and reduce heat loss.

### Fire Alarm Systems

Raise all shades to ensure visibility of emergency egress paths during evacuation events.

---

## Shading Control Strategies

### Scene-Based Control
Predefined "scenes" combine shades, lighting, and AV settings to match room usage, such as "Meeting Mode" or "Presentation Mode."

### Sensor-Driven Automation
Use daylight, temperature, or occupancy sensors to trigger shade adjustments dynamically based on real-time conditions.

### Time-of-Day Logic
Automate shade movements according to sunrise, sunset, or programmed daily schedules to optimize comfort and efficiency.

### User Overrides
Allow occupants manual control of shades, but automatically return to baseline automation after a set time period to maintain consistency.

---

## Network and Communication Considerations

Common communication pathways for shading systems include:

- **Dry Contacts or Relays**: Simple up/down commands from AV or lighting controllers.

- **RS-485 or Proprietary Buses**: Traditional wired communication for many legacy shading systems.

- **BACnet/IP or Modbus**: Increasingly common protocols for scalable commercial shading systems.

- **Wireless Mesh Networks**: Useful in retrofit projects or decentralized shading deployments.

**Design Tip:** Always verify protocol compatibility early in the design phase, especially when shades must respond to triggers from multiple systems.

---

### Coordination Tips for Successful Shading Integration

- Work closely with architects and interior designers to validate shade types, aesthetics, and mounting details.

- Ensure shade zones align logically with HVAC and lighting zones for coordinated automation.

- Document shading integration requirements clearly within Division 12 (Furnishings) and Division 25 (Integration) of the project specifications.

- Conduct daylighting and glare studies early to inform shading control strategies.

- Engage AV and lighting vendors early if they will trigger or manage shade movements.

---

### Summary

Shading and façade systems are no longer standalone accessories—they are critical elements of integrated building control.

By aligning shade automation with lighting, HVAC, AV, and occupancy systems, project teams can improve occupant comfort, reduce energy consumption, and deliver a more polished, responsive user experience.

---

**Real-World Insight – Smart Glass is Cool, but It's Not Plug-and-Play**

Dynamic smart glass systems, capable of tinting or going opaque based on light conditions, offer a sleek and futuristic alternative to traditional shading.

However, smart glass requires intensive coordination to work correctly:

- **Low-voltage wiring must be embedded inside the window mullions**, requiring early design coordination between architects, glaziers, and electricians.

- **Glass controllers must be located within a specific distance** of each pane, meaning careful planning for equipment placement and access is crucial.

- **Outdoor light sensors must be mounted on the same façade** as the smart glass they control. Placing a sensor on the wrong building face can result in inappropriate tinting or ineffective solar gain control.

Smart glass solutions are highly effective—but only when fully integrated into the building's electrical and control systems from the earliest design stages.

---

## Executive Summary

**Automated shading systems now play a critical role in building performance.**
They manage daylight exposure, reduce solar heat gain, support energy efficiency, and enhance the overall occupant experience.

**Integrated shades enhance building intelligence.**
When shading is coordinated with lighting, HVAC, AV, and occupancy systems, buildings can achieve improved comfort, operational efficiency, and aesthetics through dynamic automation.

**Successful shading integration depends on early coordination and planning.**
Key success factors include verifying protocol compatibility, aligning shading zones with other systems, documenting control strategies clearly, and ensuring that all relevant trades are engaged early in the project design process.

# Chapter 17: Audiovisual Systems — Enhancing Experience Through Integration

Audiovisual (AV) systems have become an essential part of the occupant experience in commercial buildings. Whether supporting presentations, collaboration, public messaging, or ambiance, these systems now operate as core building infrastructure.

Once siloed and highly vendor-specific, modern AV systems are increasingly integrated with lighting, shading, access control, and HVAC to create seamless, high-impact environments.

This chapter explores how AV systems operate, where they integrate, and how to ensure they enhance rather than conflict with the rest of the building's smart capabilities.

---

## Core AV System Components

### AV Processor / Controller
Manages event logic and scene execution for AV operations.

### Digital Signal Processor (DSP)
Handles audio routing, mixing, and signal processing for microphones and speakers.

### Matrix Switcher
Routes audio and video signals from various sources to selected output devices.

### Touch Panel or Wall Station

Provides a user interface to control AV scenes, source selection, and environmental presets.

### Display Device (TV or Projector)

Delivers visual output for presentations, digital signage, or media playback.

### Microphone / Speaker

Captures and outputs audio, forming the input/output core of any AV environment.

AV control systems typically operate at Purdue Model Level 2, handling local logic. Coordination with facility-wide scheduling or access control elevates them into Level 3 territory.

---

### Integration Opportunities

AV systems commonly integrate with:

- **Lighting Systems**
  Automatically adjust lighting scenes for presentations, meetings, or video calls.

- **Shading Systems**
  Lower blackout shades to improve visibility during projection or reduce glare in conference rooms.

- **HVAC Systems**
  Trigger pre-conditioning of rooms before scheduled events to maintain comfort.

- **Access Control Systems**
  Enable AV presets or unlock controls only when authorized users are present.

- **Fire Alarm Systems**
  Automatically mute microphones and reduce background audio, while displaying emergency messages on connected screens.

For deeper integration strategies, see related chapters on lighting, shading, and access control.

---

## Scene-Based Control Strategies

AV systems often operate based on "scenes"—combinations of lighting, shading, audio, and visual settings tailored to specific use cases.

A typical scene may include:

- Activating a display and selecting a video source

- Adjusting microphone and speaker volumes

- Lowering blackout shades

- Setting lighting to a predefined level

Scenes can be triggered manually, scheduled by calendar systems, or automated via occupancy detection or access control.
Priority logic should be clearly defined—emergency events must override user or scheduled actions.

## AV and Life Safety Coordination

AV systems must support emergency protocols by:

- **Audio Ducking**
  Reducing or muting background audio to prioritize public address or emergency messages.

- **Digital Signage Override**
  Replacing scheduled content with emergency information across displays.

- **System Shutdown**
  Disabling unnecessary equipment to reduce noise, distraction, or network load during life safety events.

These behaviors must comply with code and be verified during commissioning.

## Networking Considerations for AV

As AV becomes increasingly IP-based, network design must account for:

- **VLAN Assignment and Isolation**
  Segment AV traffic to avoid interference with building controls or corporate IT services.

- **Multicast Management**
  Configure switches to support AV streaming protocols.

- **Cybersecurity**
  Harden AV processors, touch panels, and servers against intrusion. Disable default passwords and monitor access logs.

- **Integration Compatibility**
  Ensure AV platforms can interface with building-wide APIs, MQTT brokers, or BACnet gateways.

Early coordination with IT is essential to prevent AV traffic from overwhelming shared infrastructure.

---

## Commissioning and Ownership

Effective AV deployment includes:

- **User Training**
  Walkthroughs for end-users on system operation, touch panels, and presets.

- **Functional Testing**
  Validate integration between AV, lighting, shading, HVAC, and life safety systems.

- **Naming and Documentation**
  Label sources, zones, and scenes clearly. Maintain editable configuration files for future updates.

- **Defined Ownership**
  Clarify long-term responsibilities between the AV integrator, IT department, and facilities team.

Deliverables should include signal path diagrams, port maps, API endpoints, and editable configuration files.

---

## A Tiered View of AV System Complexity

During my time at Waveguide, I learned a helpful model developed by Scott Walker for classifying AV system complexity:

- **Level 1** – Basic AV routing and control.

- **Level 2+** – Adds DSP, touch panels, and integrated lighting or shading.

- **Level 3–5** – Includes distributed audio, cross-room presets, and multi-zone coordination.

- **Level 6+** – High-performance environments like performing arts centers or experience centers.

In these advanced environments, AV systems coordinate tightly with HVAC airflow, lighting color temperature, structural acoustics, and more.

Even in simpler buildings, areas like executive briefing centers, digital lobbies, and innovation labs often require similar design precision.

---

## Summary

AV systems are the most visible—and often most scrutinized—technology in a building.

When well-integrated, they create seamless, intuitive environments. When siloed or misconfigured, they become frustrating and high-maintenance.

A smart building experience depends on smart AV design—with thoughtful scenes, reliable integration, and long-term support built in.

---

## Executive Summary

**AV systems define how people interact with a building.**
They support collaboration, ambiance, messaging, and branded experiences across commercial spaces.

**Integration is key.**
By coordinating AV with lighting, shading, HVAC, and access control, designers can create seamless transitions between different use cases through intelligent scene-based control.

**IP-based AV requires strong IT coordination.**
Early planning around VLANs, multicast traffic, cybersecurity, and integration protocols prevents future issues and ensures compatibility with enterprise systems.

**Commissioning matters.**
AV systems must be tested for both functionality and integration.
User training, clear documentation, and long-term ownership planning are vital for sustainable system success.

**AV is where expectations are highest.**

It's the first system occupants see—and the first they'll complain about if it fails.

Treat AV as an experience platform, not just a technology package.

# Chapter 18: Specialty Systems: Elevators, Generators, Utility Meters, and More

Not every building system fits neatly into a major silo like HVAC, lighting, or audiovisual (AV)—but that does not make them any less critical.
Elevators, emergency generators, water treatment skids, utility meters, parking systems, and digital signage all play essential roles in operations and occupant experience.

Each of these specialty systems also has significant integration potential—if they are addressed early in the project strategy.

This chapter looks at how specialty systems can be coordinated into the broader building controls ecosystem, and how to avoid the common trap of thinking, "We forgot about that one until the end."

---

**Examples of Specialty Systems and Their Integration Potential**

**Elevator and Lift Systems**

- Fire recall integration

- Occupancy triggers for HVAC or lighting adjustments

- Card access system integration for destination control

**Emergency Generator Systems**

- Monitoring run status and alarm conditions

- Tracking fuel levels for operational readiness

- Reporting generator availability to BAS or EMS dashboards

## Utility Metering

- Providing real-time consumption data for Energy Management Systems (EMS)

- Supporting demand response programs and tenant billing systems

## Parking and Gate Systems

- Integrating access control credentials for automated entry

- Displaying space availability on digital signage

## Digital Signage Systems

- Delivering emergency messaging

- Sharing energy use statistics or event information in public areas

## Irrigation and Landscape Control

- Scheduling based on weather forecasts or building occupancy patterns

## Laboratory Equipment and Pumps

- Monitoring high-level alarms, run hours, or critical status points for asset protection and environmental safety

## Photovoltaic (PV) and Wind Energy Systems

- Monitoring power generation, inverter status, and alarm conditions

- Coordinating ventilation for battery rooms or control enclosures

## Why These Systems Often Get Overlooked

- They are procured separately or late in the design process.

- Ownership may fall under departments outside of Facilities or IT, such as Security, Engineering, or Landscaping.

- They may lack standardized integration methods compared to HVAC or lighting.

- "Out of scope" excuses often arise during contract negotiations.

- Some packaged systems seem self-contained, but visibility into status and alarms still adds operational value.

## How to Integrate Specialty Systems Successfully

### Start with Visibility

- Focus on exposing alarms, statuses, or operational conditions through simple monitoring points.

### Use Gateways and Translation Tools

- Common integration protocols include BACnet, Modbus, SNMP, and MQTT.

### Map Them to Level 3 Systems

- Provide building operators a single pane of glass showing specialty system states alongside core BAS data.

### Avoid Over-Control

- For most specialty systems, monitoring is sufficient. Full command and control are unnecessary and can create additional risk.

### Respect Packaged Controls

- Allow the original system packaging to manage internal logic, especially for life-safety or critical systems like generators and elevators.

---

### Coordination Tips for Specialty System Integration

- Identify all specialty systems during early design programming and requirements gathering.

- Create an explicit checklist of these systems and plan for integration.

- Assign Division 25 scope or create specific language in system sections for vendor cooperation.

- Require specialty system vendors to cooperate with the integration team by exposing data points or supporting protocol translation.

- Confirm whether any specialty system will need to communicate across the building network—and involve IT early if so.

---

## Summary

Specialty systems are often overlooked until late in the project—but they are essential to the daily functioning and resilience of the building.
By planning for their integration early, project teams can enhance operational visibility, streamline responses to events, and maximize the value of the overall building automation strategy.

Including specialty systems in the integration scope helps ensure that elevators, generators, parking gates, and other key subsystems do not become isolated weak links within a smart building environment.

---

### Executive Summary – Chapter 18: Specialty Systems

**Specialty systems like elevators, emergency generators, utility meters, and parking systems are critical to building operations and occupant experience.**

They require intentional early planning to ensure visibility, monitoring, and appropriate integration.

**Key actions include:**

- Identifying specialty systems early and adding them to the integration checklist.

- Using gateways like BACnet, Modbus, SNMP, or MQTT to bring operational data into Level 3 systems.

- Focusing on monitoring rather than controlling specialty systems, respecting their packaged control designs.

- Confirming network requirements and securing IT coordination where needed.

- Assigning scope and requiring specialty vendors to cooperate with the integration team.

**Bottom line:**
Early coordination turns specialty systems from isolated assets into active participants in a smarter, more resilient building strategy.

# PART III: Infrastructure, Execution, and Lifecycle Management

# Chapter 19: Network Infrastructure and Structured Cabling

Behind every smart building system—HVAC, lighting, AV, access control—is a web of cables, switches, routers, and patch panels that make communication possible.

While it's easy to focus on devices and user interfaces, integration is only as strong as the infrastructure that connects those systems together.

**Structured cabling** and **network infrastructure** are both critical—but distinct—foundations of a smart building. This chapter offers an overview of structured cabling best practices, while focusing more deeply on network architecture—the area where convergence, cybersecurity, and future-proofing come together.

If you are working in a speculative, multi-tenant, or core-and-shell building, infrastructure needs differ significantly. Early consultation with property management and their IT teams is essential.

---

## What Is Structured Cabling?

Structured cabling is a standardized approach to installing and managing the physical layer of a building's technology systems.

It includes:

- Backbone cabling between telecom rooms (IDFs/MDFs)

- Horizontal cabling from telecom rooms to devices

- Patch panels, racks, and cable management hardware

- Labeling and documentation based on standards (e.g., ANSI/TIA-606)

**Important:** Structured cabling is often shared between multiple systems (BAS, AV, Security) and requires intentional coordination early in the project.

---

**The Case for Converged Networks**

Historically, each system had its own physical network:

- BAS on BACnet MS/TP or isolated IP

- Lighting on a proprietary bus

- AV on dedicated switches

- Access control managed separately by IT

Today, converged IP networks with VLAN segmentation are increasingly common.

**Advantages of a converged design include:**

- Reduced cabling and hardware costs

- Centralized cybersecurity management

- Improved interoperability between systems

- Easier maintenance and troubleshooting

**Caution:** Convergence also introduces new demands for bandwidth, quality of service (QoS), and cybersecurity. IT must be actively involved from day one.

---

### Estimated Bandwidth Planning

Building systems vary dramatically in their network demands.

- **Video Surveillance** and **AV Systems** are typically the largest consumers of bandwidth.

- **BAS**, **Lighting**, and **Access Control** have smaller, but critical, consistent traffic needs.

**Design Tip:** Plan network capacity based on real traffic models, not just device counts. Work with vendors early to estimate bandwidth per system.

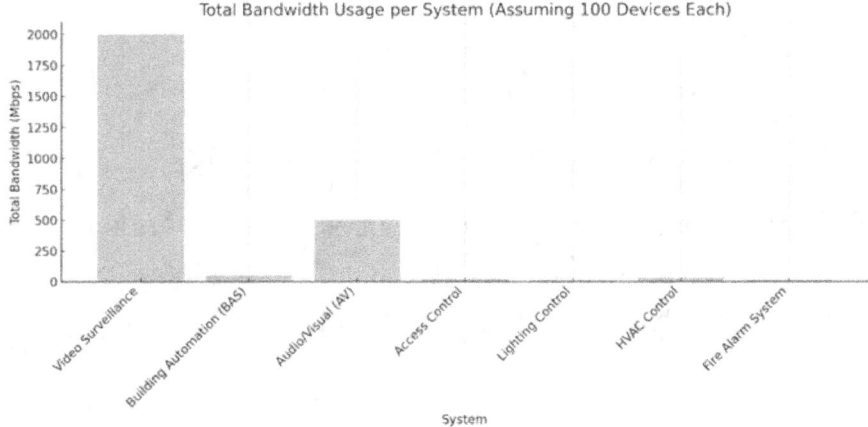

**Figure: Estimated Bandwidth Usage by System (Assuming 100 Devices Each)**

This chart compares total network bandwidth consumption for different systems.

- **Video Surveillance** leads significantly, consuming over 2000 Mbps

- **Audiovisual (AV)** systems follow, at approximately 500 Mbps

- **BAS**, **Access Control**, and **HVAC** use far less bandwidth, typically under 100 Mbps

- **Lighting** and **Fire Alarm Systems** register negligible bandwidth loads

*(See image below for comparison of usage across systems)*

---

### VLAN Design Best Practices

Virtual LANs (VLANs) allow segmentation of traffic on a shared physical network.

**Best practices include:**

- Create separate VLANs for each major system (BAS, AV, Access Control, Lighting).

- Assign static IP ranges and subnet masks for consistency.

- Restrict cross-VLAN access with firewalls or access control lists (ACLs).

- Implement QoS to prioritize latency-sensitive traffic (especially AV and real-time control systems).

**From experience:** Involving IT early in VLAN and IP scheme design prevents 90% of downstream coordination issues.

---

### Common Cabling Types and Pathways

### Cat6 / Cat6A

- Standard for Ethernet device drops (IP cameras, controllers, sensors).

### Fiber Optic

- Backbone links between telecom rooms (IDFs) to handle high bandwidth over long distances.

### RS-485 (Twisted Pair)

- Legacy fieldbus wiring for BACnet MS/TP, Modbus RTU, and some lighting systems.

### Shielded Audio Cable

- Used for sensitive low-voltage audio signals in AV systems.

### Low-Voltage Control Wiring

- Dry contact or relay connections for simple binary signals.

**Design Tip:** Document cable pathways and pull schedules early. Coordinate closely between electrical, IT, and specialty contractors.

---

## Color Coding for Future-Proofing and Cybersecurity

Color-coding network cables by system type or function dramatically improves:

- Field identification

- Troubleshooting speed

- Cybersecurity awareness

- Risk mitigation against cross-connection errors

Color-coding costs pennies if planned early. Document the color standard in the design drawings, specifications, and turnover documentation.

---

## Infrastructure Coordination Essentials

- Reserve IDF/MDF rack and wall space early—demand always exceeds initial estimates.

- Confirm dedicated power circuits, UPS needs, and ventilation for network closets.

- Use common cable trays and conduit paths wherever possible.

- Enforce strict labeling standards across all trades.

- Include cable certification test results and updated as-built drawings in closeout deliverables.

---

## Red Flags to Watch For

Several warning signs can indicate future risk in building infrastructure:

- Devices using **proprietary protocols** without clear segmentation, encryption, or monitoring support.

- Vendors requiring **open inbound ports** without secure architecture justifications.

- Hardcoded IP addressing or unmanaged broadcast traffic across the LAN.

- Systems lacking documented **port maps**, **APIs**, or **vulnerability disclosures** (including a Software Bill of Materials, or SBOM).

- Cloud-first platforms without local failover or clear data sovereignty policies.

- Licensing models that inhibit growth (e.g., per-point fees or per-device cloud fees without scaling plans).

**Bottom line:** Visibility, segmentation, scalability, and maintainability must be non-negotiable features of any building technology deployment.

---

## Summary

A building's smart strategy is only as strong as its infrastructure.

Thoughtful planning of structured cabling and network architecture:

- Simplifies maintenance

- Improves cybersecurity and uptime

- Enables easier integration of new technologies

- Future-proofs the building against evolving tenant, regulatory, and operational demands

Done right, the infrastructure supports not just today's integration goals—but the next decade of innovation.

---

**Executive Summary**

**Smart buildings depend on strong infrastructure foundations.**
Both structured cabling and resilient IP networks are

essential—but they are distinct disciplines requiring careful coordination.

**Key Takeaways:**

- **Structured cabling** is the physical backbone—consistent standards for routing, labeling, and documentation matter.

- **Converged networks** with VLAN segmentation offer scalability and cybersecurity—but only succeed when coordinated closely with IT.

- **Red flags** like unmanaged addressing, open inbound ports, or non-cooperative vendors must be addressed before commissioning.

- **Infrastructure is a strategic asset**—not just a technical requirement.
  It enables secure, flexible, insight-ready buildings today and into the future.

# Chapter 20: Integration's Final Test: Handoff Done Right

No matter how advanced your building systems are, if they aren't documented, tested, and handed over properly, the project—and the people who operate it—will struggle.
The **final 5% of the project determines 95% of the owner's perception**.

Integration lives and dies in the handoff.

---

## Why Documentation Matters

Good documentation isn't red tape—it's the DNA of the system. It captures:

- What was installed
- How it was configured
- Who owns and maintains it
- How it's supposed to behave

A complete turnover set should include:

- As-built control drawings
- Network diagrams with VLANs, IPs, and ports
- Point lists and naming standards
- Updated sequences of operation
- Integration matrices

- Panel and termination schedules

- User manuals and training content

- Credential turnover logs and password storage practices

But documentation can't be static.
It must become a **living record** of the building—tracking not just the *as-built*, but the *as-operated*.

Just like source control in software, operators need to know:
What changed, who changed it, when, why—and what happened next.

---

### Real-World Contrast

**When documentation fails:**
I once had to reverse-engineer an entire system—reading legacy controller code line-by-line, physically tracing wires through conduit—because nothing was labeled or saved.

**When it works:**
A job where every wire had printed labels matching point names, drawings aligned with the graphics, and the turnover package included GUI mockups and test scripts. That building ran clean from day one.

---

### Commissioning and Testing

**Commissioning is where design meets reality.**
This is often where integration failures reveal themselves.

You need:

- Functional test plans for each system
- Contractor Field Tests completed before the commissioning agent arrives
- Integration point verifications
- Alarm and trend validation
- GUI and graphics usability checks
- IT review of network segmentation and cybersecurity measures

**Also include an Integrated Systems Test (IST)**—a scripted walkthrough of real-world scenarios:

- Fire alarm triggers
- Occupant badge access
- Scheduled AV scenes
- Power outages and restarts

This isn't just QA.
It's storytelling. Show how the building will behave when it matters.

---

## Owner Training

Training is not optional.
A smart building without trained operators is a liability.

Provide:

- Live walkthroughs for IT and facilities teams

- "Train-the-trainer" sessions for internal champions

- Video tutorials and how-to sheets

- One-page reference guides for common workflows

- Escalation paths for support

**Start training plans as early as design.**
No one should be surprised by how the system works
during training.

---

**Turnover Package Essentials**

Don't deliver a stack of PDFs and call it a day.
Provide an organized, digital package that includes:

- Editable drawings, sequences, and graphics

- Passwords and credential reset protocols

- Warranty and asset tag lists

- Preventive maintenance checklists

- Integration-ready files for CMMS or digital twin
  systems

Organize it in clearly named folders (e.g., "01 Drawings", "02 Sequences", "03 Training") and host it on a shared cloud location (e.g., SharePoint, Egnyte, Procore).

---

**Long-Term Maintenance Planning**

Integration isn't "set it and forget it."

Make the system sustainable:

- Schedule regular patching and firmware updates

- Rotate credentials and track expiration

- Validate remote access protocols

- Review alarm and trend logs periodically

- Test backups and restoration processes

Assign clear roles:
Who updates firmware? Who changes passwords? Who handles backups?

Make sure your plan can live inside the owner's CMMS or digital twin ecosystem.

---

**Final Thought**

**The success of your integration effort isn't measured on Day 1—it's measured on Day 1,000.**
If you hand over a building with no roadmap for operation, no one will trust the system.

But when documentation is strong, testing is thorough, and the owner feels empowered?

That's when the building begins to thrive.

"Nobody cares how you fly the plane if you can't land it."

---

## Executive Summary

- **Final impressions last:** The last 5% of work—documentation, testing, training—shapes 95% of the owner's experience.

- **Documentation must live:** Track changes over time. A static drawing set isn't enough.

- **Test end-to-end:** Simulate real-world scenarios during commissioning, not just device checks.

- **Train for success:** Prepare users, IT, and internal champions with workflows—not just features.

- **Deliver digital packages:** Include editable drawings, logic files, and operational instructions in cloud-based handover tools.

- **Sustain the system:** Assign roles and schedules for updates, audits, and backups.

# Chapter 21: Maintenance, Upgrades, and Lifecycle Planning

Even the best-integrated building will fail if it's not maintained.

Smart systems aren't just about **Day One success**—they're about staying functional, secure, and valuable for **years**.
That takes a plan.
This chapter shows you how to create one.

---

### Why Lifecycle Planning Matters

Building systems don't stay new.

- Controllers reach end-of-life

- Software becomes unsupported

- Network standards evolve

- Staff turns over

- Vendors disappear

**If there's no plan, the system decays.**
Integrations drift. Vulnerabilities grow. Costs spike.

Lifecycle planning is the long-term solution.

---

### Core Maintenance Activities

Set a recurring calendar and assign owners:

- **Backup checks** – Quarterly

- **Remote access test** – Semi-annually

- **Alarm & trend log reviews** – Monthly

- **Firmware/software validation** – Annually

- **Battery/sensor replacement** – Based on device specs

- **User access/role review** – Bi-annually or after staff change

Don't rely on memory. Make this a **repeatable checklist**.

---

## Firmware and Software Upgrades

You can't ignore updates. But you **can** do them right.

- Track all versions in a central log

- Schedule during low-occupancy times

- Test on a sandbox system if possible

- Backup *before* updating

- Document what changed and why

Always involve IT before touching the network.

---

## Monitoring and Analytics

Level 3 and Level 4 platforms can do more than display temperatures.

Use analytics to:

- Flag repeated overrides or overrides left in place
- Detect underperforming sequences
- Spot energy waste or occupant discomfort
- Benchmark zones and systems for planning

Let the building tell you what it needs. Then act on it.

---

## Plan for Replacement—Not Surprise

Everything fails eventually. But not everything has to be a surprise.

Build a forecast model:

- **Controllers** – 10–15 years
- **Network switches** – 5–7 years
- **Software platforms** – Varies; check vendor EOL
- **Cloud services** – Watch licensing changes and sunset notices

Budget 3, 5, and 10 years out.

---

## Managing Change and Growth

Smart buildings evolve. So should your systems.

Document your process for:

- Adding or modifying zones
- Reconfiguring lighting or HVAC
- Seasonal schedule changes
- Integrating new subsystems
- Swapping equipment while maintaining integration

Use **change logs** to track what's been done—and why.

---

### 🔧 Real-World Insight: Still Standing After 30 Years

I walked back onto a site I helped commission in **1995**. It's now 2025—and still going.

It wasn't cutting-edge, but it was **reliable**. Why?
Because every **first Monday** of the month, the team does a full system walk:

- They test everything
- They check sequences
- They inspect panels

That system is alive because someone made it their job to keep it that way.
**Integration isn't just about tech—it's about stewardship.**

---

## Summary

A smart building doesn't stay smart by accident.
It takes structure, ownership, and intentional care.

Lifecycle planning transforms integration from a **project phase** into a **sustainable mindset**.

---

## Executive Summary

- **Smart systems need upkeep:** Maintenance, testing, and upgrades keep systems aligned and secure.

- **Without a plan, systems drift:** Gaps emerge, vulnerabilities grow, and integration fails silently.

- **Maintenance must be routine:** Schedule backups, firmware checks, access reviews, and log audits.

- **Upgrades need discipline:** Always test first, coordinate with IT, and document everything.

- **Analytics = insight:** Use override flags, energy drift, and trend analysis to drive action.

- **Expect replacement:** Budget for controller and platform refreshes based on service life.

- **Track changes:** Use logs and processes to manage system evolution and operational growth.

Integration is a lifecycle—not a one-time effort.
Plan for the long haul, and your building will repay that effort every day.

# Chapter 22: Making the Business Case for Integration

**Integration doesn't sell itself.**

But it doesn't need a sales pitch—it needs a **translator**.

Most building owners, contractors, and executives aren't tech experts. They're focused on budgets, timelines, outcomes.

So, our job isn't to convince them that integration is *possible*—our job is to show them it's **valuable**.

This chapter is your cheat sheet for doing just that.

---

## What You *Really* Do When You Pitch

Looking back, I don't remember giving a presentation about integration.
I remember listening.

- The GC was worried about delays.

- The owner didn't want another 2am support call.

- The exec wanted tenant satisfaction.

**I didn't pitch a product. I told them a story.**
A story about how we solved that same problem last time.

Decision-makers aren't asking: *Can it be done?*
They're asking:

- Will it work?

- Will it cause new problems?

- Can my team support it?

- Will this blow up my schedule or budget?

If you answer those questions with **empathy and clarity**, you're already halfway there.

---

## Know Your Audience

Different stakeholders want different things. Match your message.

### Primary Concerns by Role

### General Contractor:
Focuses on schedule, scope clarity, and coordination.

---

### Executive / C-Suite:
Concerned with return on investment (ROI), reputation, and tenant experience.

---

### Facility Manager:
Prioritizes reliability, maintenance needs, and energy efficiency.

---

### IT Director:
Focuses on cybersecurity and long-term supportability.

**Architect / Engineer:**

Ensures design intent is preserved and code compliance is achieved.

One pitch doesn't fit all. Tailor your message.

## The Real Value of Integration

Integration isn't about features. It's about outcomes:

- Fewer callbacks and rework

- Faster commissioning

- Lower energy bills

- Unified platforms = fewer logins

- Smarter alarms and analytics

- Happier tenants

- Fewer 2am phone calls

Tell stories. Share examples. Use their vocabulary.

## Speak in Dollars, Days, and Frustrations

Bad:

"This streamlines your control system via multi-protocol support."

Good:

"This keeps your team from spending hours trying to get one system to talk to another."

Great:

"You'll save 8–12% on utility costs annually—just by coordinating HVAC and lighting schedules."

Make it concrete. Make it real.

---

## A Framework You Already Use

You don't need a slide deck—you need this mental model:

**Start with the pain.** "Last time, commissioning dragged out because lighting and HVAC weren't coordinated..."

**Explain how integration solves it.** "If we set those sequences now, you avoid all that rework later."

**Close with outcomes.** "We've cut commissioning time by 20% using this strategy. Plus, support calls drop when systems are visible in one place."

If you're talking to the GC:

"This keeps your electrician from crawling into the same ceiling four times."

---

**Timing the Ask**

Best time to talk integration?
**Before**:

- MEP drawings are finalized

- Trades are selected

- Network architecture is locked in

**Early = leverage.**
**Late = change orders.**

---

**What to Keep in Your Toolkit**

- A real project example

- One or two data points (energy, time, cost saved)

- A **plain English** explanation of what integration means *on this job*

- A follow-up question that moves the conversation forward

---

**Don't Forget to Make the Ask**

Too many engineers explain—but forget to ask.

Try:

- "Can we include this in early coordination scope?"

- "Should we reserve rack space now to avoid a retrofit later?"
- "Would it help to schedule a session to map what systems need to talk?"

**You're not closing a deal. You're inviting a decision.**

---

## Objections and How to Handle Them

When you propose integration, expect pushback. That's normal. Your job isn't to argue—it's to listen, empathize, and reframe the conversation.

Here are some common objections and how to respond:

**Objection:** "We've always done it this way."
**Response:** "And that way doesn't scale. Let's talk about how things are changing—and how we can stay ahead of it."

**Objection:** "It costs too much."
**Response:** "Let's model what it saves over ten years, not just on day one. Integration pays off when you look at lifecycle value."

**Objection:** "We don't have in-house skills."
**Response:** "We can support your team—and train them as we go. You don't have to go it alone."

**Objection:** "It'll slow the project down."
**Response:** "Done right, integration reduces rework, prevents change orders, and makes commissioning smoother. It can actually save time."

## Summary

You don't need to sell integration.
You need to **explain it like a human.** With empathy. With proof. With a clear next step.

That's how integration wins—not in the specs, but in the **conversation**.

---

### 📷 Executive Summary

- **Integration isn't sold—it's translated.** Speak to outcomes, not protocols.

- **Match your message to the role:** GCs want schedule certainty. Execs want ROI. IT wants security. Ops teams want reliability.

- **Use plain English:** Tell real stories. Quote real numbers.

- **Make the ask early:** Before design is fixed. Before trades are hired.

- **Keep your toolkit simple:** A story, a stat, and a follow-up question.

- **Objections are normal:** Respond with empathy and a better path forward.

If you want integration to succeed—don't just install it. **Advocate for it.**

# Chapter 23: Standards, Codes, and Specifications

Smart building integration may feel like the Wild West at times, but there's a solid legal and technical backbone behind it—if you know where to look. From cybersecurity to interoperability, a growing number of standards and guidelines can help you align your project with best practices, reduce risk, and gain credibility when working with engineers, contractors, and IT.

This chapter reviews the key frameworks, specifications, and industry documents that support integrated design and operation.

Here's a **structured outline** of common and recognized **Division 25 sections**, based on CSI MasterFormat and expanded industry usage:

---

## Division 25 – Integrated Automation

### 25 00 00 – General

- 25 00 10 Integrated Automation General Requirements

---

### 25 05 00 – Common Work Results for Integrated Automation

- 25 05 01 System Integration Coordination (*custom or simplified spec section*)

- 25 05 10 Network Infrastructure for Integrated Automation

- 25 05 11 Cybersecurity for Facility-Related Control Systems (aligns with UFGS)

- 25 05 13 System Interfaces for Integrated Automation

- 25 05 14 Integrated Automation Communication Protocols

- 25 05 15 Integrated Automation Software and Licensing Requirements

- 25 05 16 Integrated Automation Field Equipment Panels

- 25 05 17 Integrated Automation Signal Types and Terminations

---

## 25 10 00 – Integrated Automation Network Equipment

- 25 10 10 BAS/UMCS Front-End and Integration

- 25 10 20 Supervisory Controllers and Gateways

- 25 10 30 Network Switches, Routers, and Firewalls

- 25 10 40 System Time Synchronization and Time Servers

---

## 25 30 00 – Integrated Automation Instrumentation and Terminal Devices

- 25 30 10 Sensors for Temperature, Humidity, CO2, etc.

- 25 30 20 Input/Output Modules and Terminal Strips

- 25 30 30 Device Addressing and Identification

---

## 25 50 00 – Integrated Automation Facility Management (FM) Interfaces

- 25 50 10 Interfaces to Lighting Control Systems

- 25 50 20 Interfaces to HVAC Control Systems

- 25 50 30 Interfaces to Fire Alarm Systems

- 25 50 40 Interfaces to Access Control and Security

- 25 50 50 Interfaces to Audio Visual and Paging Systems

- 25 50 60 Interfaces to Utility Metering / Smart Grid Systems

- 25 50 70 Interfaces to Elevators and Vertical Transport

---

## 25 90 00 – Integrated Automation Control Sequences

- 25 90 10 Sequence of Operations – HVAC Integration

- 25 90 20 Sequence of Operations – Lighting Integration

- 25 90 30 Sequence of Operations – Life Safety Integration

- 25 90 40 Sequence of Operations – Energy and Demand Management

---

**Additional Common Custom Sections (Non-CSI Standard but Industry-Used)**

- 25 99 00 – Integration Testing and Commissioning

- 25 99 10 – Integration Matrix

- 25 99 20 – Owner Training and Handover

A complete sample specification is included in Appendix F for reference.

When trying to decide how to do things though we really want to follow a standard. If the industry has agreed this is the minimum requirement for a type of work, you need to make sure you above that minimum.

Start with cybersecurity standards. Much like a building project, the locks are in the same paragraph as the doors. It may not describe them in as much detail but they are included at the outset. Cybersecurity should be the same way. It is the locks on your system, mention it as soon as you start a project. Here are some of the primary Cybersecurity Standards that should be on your quick reference list.

---

## Cybersecurity Standards

Here are key cybersecurity standards you should be familiar with when designing or specifying building automation systems:

- **NIST SP 800-82:** This is the U.S. National Institute of Standards and Technology's guideline for securing Industrial Control Systems. It's a comprehensive, practical reference for OT cybersecurity planning.

- **IEC 62443:** A globally recognized standard focused specifically on cybersecurity for industrial automation and control systems. It's widely accepted in international and multi-site projects.

- **UL 2900 Series:** This set of standards provides third-party cybersecurity certification for connected products and devices. It validates that systems meet a minimum security threshold.

- **CSI MasterFormat Section 25 05 11:** This is a Division 25 specification section titled *Cybersecurity for Facility Controls*. It helps integrate cybersecurity requirements directly into construction documents and project scopes.

Use these standards to align expectations across design teams, vendors, and IT departments—and to ensure your control system is secure by design, not by assumption.

Beyond cybersecurity standards but before the different silos of automation are the standards for communication between the different systems. These communication standards should be agreed to as early as possible, and with as much definition as possible. Do not think though that a system will only use one of these. Typically there will need to be at least one standard per Level of the system, and likely more than that. Standards also need to consider the silo of the system they are working in. AV communication standards are not suitable for HVAC control, but they may agree on a standard to share information between them. It may be a different standard though to go from the Fire Alarm to the HVAC system, and it's almost guaranteed it will. Starting with the most restrictive silo and working out can be an effective way to either minimize the standards or to build out a robust plan.

### Integration and Control Standards

When specifying or evaluating building systems for integration, understanding common control and communication standards is essential. Here are some of the most widely used:

- **ASHRAE 135 (BACnet):** A foundational open communication protocol used primarily in HVAC systems but also adopted by lighting and building automation platforms. It allows for vendor-neutral integration across building systems.

- **Modbus (TCP and RTU):** A robust and widely adopted industrial communication standard, especially common in utility meters, generators, and packaged equipment. Simple and effective, especially for read-only data.

- **KNX, DALI, and DMX:** These protocols are common in lighting and shade control. KNX is typically used in European systems for building control. DALI specializes in dimming and fixture-level lighting control. DMX is used primarily for theatrical or display lighting, especially in museums and auditoriums.

- **MQTT and REST APIs:** These are modern, lightweight integration tools widely used in cloud applications, analytics platforms, and enterprise dashboards. They are ideal for real-time messaging and data exchange between IT and OT systems.

**Why it matters:**
Choosing systems that support open protocols provides long-term flexibility, eases future upgrades, and helps prevent vendor lock-in.

---

Standards can also dictate how systems are designed or operated. Again, these are minimum requirements for the industry. Failure to meet these requirements can have a negative impact on your project. Here are some basic standards for the different silos of automation in a building:

**Commissioning and Performance Guidelines**

To ensure building systems are installed, integrated, and verified properly, it's critical to reference established commissioning and performance standards. The following guidelines offer structure, clarity, and industry consensus for evaluating both control systems and their operational outcomes:

- **ASHRAE Guideline 13:** This guideline provides best practices for specifying control systems in building projects. It helps ensure that design documents and specifications clearly define system functionality, sequences, and integration requirements.

- **ASHRAE Guideline 0:** Known as the Total Building Commissioning Process, this guideline outlines the framework for planning, executing, and documenting a comprehensive commissioning effort across all systems in a facility.

- **ACG / BCA / NEBB Commissioning Forms:** These are structured checklists and forms used by commissioning professionals from organizations like the AABC Commissioning Group (ACG), Building Commissioning Association (BCA), and National Environmental Balancing Bureau (NEBB). They support consistent testing, verification, and reporting of building system performance.

- **ANSI/AVIXA V201.01:2021:** This is a standard focused on audiovisual systems and energy management. It addresses how AV systems should be designed and operated for optimal energy use, particularly in integrated environments.

**Why it matters:**
Using commissioning guidelines ensures systems are not only installed—but tested, verified, and documented. Integration success depends heavily on these practices during handover and long-term operation.

These help ensure integrated systems are delivered, verified, and documented properly. If there are solid and defensible reasons for not using a standard though those need to be included in the specifications for the project. An example that comes to my mind is a winery project that a colleague of mine completed had a patented ventilation system in order to improve the wine making process. This sequence of operation was completely different than any industry standard, and had to be explained to the Authority Having Jurisdiction (AHJ) for approval. But it still had to comply with safety standards for fire reasons.

---

## Typical Division Sections and Their Scope

### Section 26 09 23
Lighting control devices (sometimes integrated via Division 25)

---

### Section 27 15 00
Structured cabling

## Section 28 13 00
Access control

## Section 12 24 00
Motorized shades

## Section 33 49 00
Site-level systems like irrigation and utility metering

Cite these in your specs to avoid scope gaps or conflicting responsibilities. These sections of the specs will often have automation devices that need to be integrated but are commonly siloed and forgotten until it's too late. See the appendix for an example of a Division 25 specification that references other portions of the project specifications.

Take for example the irrigation system. This system will benefit from having access to the HVAC systems weather station or weather forecast feed. But irrigation is located outside of the main building, and conduit paths can be prohibitively expensive to add late in construction. The irrigation system may also need to wired up in a specific manner to be able to control it remotely versus using some internal timers.

Going a step further though, what is that system integration intended to do? Is it simply to turn on the system at optimum times for water conservation? Or does it cover something like an internal courtyard, and that space needs to be controlled to allow it's use for some corporate events?

---

How do you learn all of this? Or better yet, help raise the bar for the entire industry and set a new minimum standard.

### Supporting Organizations & References

To strengthen your specifications, designs, and business cases, it's helpful to reference respected organizations within the built environment and automation industries. These groups provide standards, white papers, certifications, and ongoing research that can validate your approach.

- **CABA (Continental Automated Buildings Association):** Focuses on smart building research and publishes detailed white papers that offer insight into trends, technologies, and return on investment for building systems.

- **IBCon / Realcomm:** These conferences and affiliated resources provide thought leadership, case studies, and best practices from across the smart building and PropTech sectors. Their materials are widely respected by owners, IT leaders, and consultants.

- **ISA (International Society of Automation):** Develops and maintains standards for industrial automation and control systems, including cybersecurity and integration protocols relevant to building systems.

- **AIA / CSI (American Institute of Architects / Construction Specifications Institute):** These bodies shape how projects are designed and specified, including Division 25 (Integrated Automation) and other MasterFormat sections critical to multi-trade coordination.

- **AVIXA (Audiovisual and Integrated Experience Association):** The standards body and trade group for AV professionals. AVIXA produces guidelines that help coordinate audiovisual systems within integrated environments.

- **SIA (Security Industry Association):** Focuses on commercial electronic security systems. Their content can help justify access control, video surveillance, and intrusion detection integrations.

**Tip:**
When writing proposals or project documentation, referencing guidance from these organizations can help establish authority, reduce risk, and align your project with accepted industry practices.

---

## Summary

You don't have to go it alone. There's a rich ecosystem of standards, guidelines, and specs that support smart building integration. Learn them, cite them, and use them to build better projects with less friction.

Up next: It's time to pull it all together. Let's walk through a sample integrated project from kickoff to closeout.

**Executive Summary**

- **Integration isn't chaos—it just feels like it without a standard.**
  Codes and frameworks bring clarity and consistency across trades and silos.

- **Start with Division 25.**
  Use CSI MasterFormat early to scope, specify, and coordinate building systems cleanly and effectively.

- **Cybersecurity must be foundational.**
  Reference NIST SP 800-82, IEC 62443, and UL 2900 from the very beginning—cyber is your building's digital door lock.

- **Agree on communication protocols early.**
  Choose protocols based on system level (e.g., BACnet for HVAC, MQTT for cloud integrations)—not based on vendor preference.

- **Use system-specific standards.**
  Each major system (AV, HVAC, fire alarm, lighting) has its own "spec language." Learn them, align them, and reference them properly.

- **Specs are only useful if they connect.**
  Division 25 must connect tightly to Divisions 26

(Electrical), 27 (Communications), 28 (Security), and others to close integration gaps.

- **Open doesn't always mean free.**
  Platforms like Niagara support open protocols, but still require licensed tools, certified integrators, and careful project coordination.

---

**Bottom line:**
Standards are your blueprint for clarity, continuity, and control. Use them to reduce friction, raise confidence, and build smarter—from design through turnover.

## Explainer: What Is a Standard—and Why Does It Matter?

### What Is a Standard?

A standard is a documented agreement established by a recognized body that provides:

- Requirements

- Guidelines

- Specifications

- Best practices

Standards ensure consistency in how systems are designed, built, operated, and maintained. They can be:

- Open (such as BACnet or Modbus)

- Regulatory (such as NFPA 72 or UL 864)

- Proprietary (such as Niagara Framework or LonMark Profiles)

- Voluntary or Mandatory, depending on jurisdiction and project scope

---

### Why Follow a Standard?

Following standards brings major benefits:

- **Interoperability:** Different systems and vendors can communicate reliably.

- **Repeatability:** Clarifies expectations and improves consistency across projects and teams.

- **Scalability:** Allows future additions and upgrades without redoing the entire system.

- **Cost Efficiency:** Minimizes custom programming and avoids vendor lock-in.

- **Compliance:** Aligns with codes, insurance requirements, and regulatory inspections.

- **Quality Assurance:** Ensures systems meet known benchmarks for performance and safety.

---

**Real-World Examples of Standards at Work:**

- **ASHRAE 135 (BACnet):** Ensures open communication between BAS devices.

- **NFPA 72:** Establishes code-compliant fire alarm response and signaling.

- **UL 864:** Governs listed fire alarm control panels for product certification.

- **IEEE 802.3:** Standardizes network cabling and switch compatibility.

- **ISA 62443:** Guides cybersecurity in automation and industrial control systems.

---

**Bottom Line:**

Standards bring order to complexity.

They create a shared language across trades and disciplines, helping diverse teams work toward a predictable, maintainable outcome.

---

## Sidebar Explainer: What Makes Niagara Proprietary?

The Niagara Framework by Tridium is often described as an "open integration platform" because it supports open protocols like BACnet, Modbus, and KNX. However, it remains a proprietary system because:

- It is owned and licensed by Tridium (Honeywell).

- Configuration and programming require Niagara Workbench and certified tools.

- Integration drivers are digitally signed and restricted to licensed developers.

- There is no open-source access to modify or redistribute its code.

- Long-term maintenance usually requires Niagara-certified integrators.

Despite its proprietary nature, Niagara remains extremely popular for cross-system integration due to its protocol support, stability, and global network of trained professionals.

Use Niagara when you need robust protocol bridging and a

standardized front-end—but plan for the licensing and support costs.

---

## Sidebar Comparison: Niagara vs. Open-Source BAS Platforms

### Niagara Framework:

- Proprietary licensing under Tridium.

- Supports BACnet, Modbus, KNX, and oBIX via licensed drivers.

- Developer access limited to certified partners.

- Custom logic developed via Java and Tridium SDK.

- Ideal for enterprise-grade, commercial BAS installations.

### Sedona Framework:

- Open-source under a BSD-style license.

- Supports Modbus and BACnet with community-driven extensions.

- Available to all developers.

- Drag-and-drop programming using Sedona Workbench or FoxTool.

- Ideal for lightweight edge control applications.

### Node-RED and Custom Open Source:

- Open-source under Apache, MIT, or similar licenses.

- Supports MQTT, Modbus, BACnet via plugins and community nodes.

- Open to all developers.

- Visual programming based on JavaScript flows.

- Ideal for experimental projects, IT/OT bridging, and rapid dashboard development.

**Takeaway:**
Open-source platforms offer flexibility and cost savings, but typically require more integration effort and support planning.
Niagara excels in tightly managed environments, offering powerful stability at the cost of flexibility and licensing control.

# Chapter 24: An Integrated Project from Kickoff to Closeout

You have seen the systems, the standards, the specs, and the strategies—now let's put it all together. This chapter walks through a hypothetical integrated building project, highlighting decisions, pitfalls, and integration opportunities at every step. It is not a fantasy project. It is a real-world, slightly messy, always evolving case study of how integrated buildings actually get built.

## Phase 1: Programming and Design Kickoff

Key Actions:

- Define integration goals with the owner early.

- Identify all relevant systems and stakeholders.

- Engage IT and cybersecurity representatives before design is locked.

- Begin initial integration diagrams and network architecture.

Typical Challenges:

- Overlooking specialty systems or retrofits.

- Owner unsure of long-term support plans.

- Each trade proposing its own siloed front-end.

Pro Tip: Write Division 25 early. Do not wait for "someone else" to define integration.

**Phase 2: Design Development**

Key Actions:

- Finalize integration platform strategy (cloud vs. on-premises, open vs. proprietary).

- Assign VLANs and IP ranges for building systems.

- Coordinate structured cabling with Division 27.

- Confirm point lists and sequences for each system.

Typical Challenges:

- Scope gaps between divisions such as Division 25 versus 23 or 26.

- Unclear expectations for ownership after turnover.

Pro Tip: Host a design coordination workshop focused solely on system integration.

**Phase 3: Procurement and Preconstruction**

Key Actions:

- Align contracts with the specifications and require cooperation with integration.

- Select vendors with proven multi-system integration experience.

- Review control panel shop drawings and integration points.

Typical Challenges:

- Vendors attempting to "value engineer" integration gateways out of the project.

- Poor documentation from manufacturers.

Pro Tip: Require submittals for integration interfaces, including BACnet tables, APIs, and network requirements.

## Phase 4: Construction and Installation

Key Actions:

- Monitor cabling and terminations for adherence to specifications.

- Ensure field devices are labeled and located per plans.

- Start early testing of system communication. Do not wait for final commissioning.

Typical Challenges:

- Network conflicts or blocked ports.

- Improvised wiring changes without updates to documentation.

Pro Tip: Commission each system individually first, then test integration across systems in phases.

## Phase 5: Integrated Systems Testing and Turnover

Key Actions:

- Run a structured Integrated Systems Test with scripts covering fire alarm events, lighting scenes, AV coordination, and other use cases.

- Deliver all documentation including as-builts, naming standards, and user roles.

- Train operations and IT staff on ongoing maintenance.

Typical Challenges:

- Incomplete training or failure to test real-world failure scenarios.

- Turnover packages missing login credentials or network maps.

Pro Tip: Host a "live-fire" dry run with all disciplines before the official handoff.

## Phase 6: Post-Occupancy Support

Key Actions:

- Set up monitoring dashboards and ticket workflows.

- Schedule 30, 60, and 90 day follow-up site visits.

- Plan for patching, backups, and version tracking.

Typical Challenges:

- Changes made post-handoff with no documentation trail.

- Staff turnover leading to loss of integration knowledge.

Pro Tip: Offer ongoing service plans or managed integration support if the facility team needs help staying current.

## Summary

Integration is not just a concept. It is a process. When planned, tracked, and executed through each project phase, integration becomes more than a technical feature. It becomes the operational foundation of the building.

## Executive Summary

- Integration succeeds when it is treated as a process, not a feature.

- Programming matters. Define goals, involve IT early, and identify all systems, not just MEP.

- Design matters. Lock in sequences, network needs, and scope clarity across divisions.

- Procurement matters. Vet vendors for experience and enforce integration compliance through submittals.

- Construction matters. Enforce labeling, cabling, and early system testing. Do not wait for final commissioning.

- Testing and Turnover matter. Integrated Systems Tests must reflect real scenarios. Deliver clean documents and hands-on training.

- Post-Occupancy matters. Plan for patches, version control, and futureproofing through support workflows.

- Practical moves throughout the process reduce friction and protect project outcomes.

Bottom line: Integrated buildings are built phase by phase. The earlier you plan, the smoother you finish.

# Chapter 25: The Future-Ready Building

A future-ready building is not just one with the latest technology. It is one that is prepared to evolve. It is designed with flexibility, built on standards, and managed with foresight. More importantly, it reflects a shift in how we think about buildings. They are not static infrastructure, but dynamic platforms for occupant experience, operational intelligence, and business value.

This chapter looks ahead. What does a future-ready building look like? How do we prepare for technologies that have not been invented yet? And what role do you play in making it happen?

**Characteristics of a Future-Ready Building**

- Interoperable Systems: Avoid vendor lock-in and allow flexible upgrades.

- Scalable Infrastructure: Prepare for more devices, bandwidth, and complexity.

- Strong Cybersecurity: Secure data, continuity, and remote access.

- Open Protocols: Use standards like BACnet, MQTT, and REST for integration.

- Digital Twin Capabilities: Support real-time analytics and modeling.

- Lifecycle Documentation: Ensure continuity across turnover and staffing changes.

- Managed Change: Build-in processes for adding and modifying systems.

Future-ready does not mean futuristic. It means resilient, adaptable, and intelligent.

**From Reactive to Proactive**

Today, most buildings are still managed reactively. Fix the alarm after it happens. Adjust the system once someone complains. Replace the controller when it dies.

Tomorrow's facilities will be different. They will predict failures through analytics. They will respond automatically with coordinated logic. They will track changes and provide transparency to stakeholders. They will serve as assets to the organization, not liabilities.

Integration is the foundation of that future.

**How to Stay Relevant**

Whether you are a contractor, engineer, facility manager, or integrator, staying relevant means:

- Learning how to talk to IT, not just trade partners.

- Understanding the language of business and ROI.

- Staying current with standards like IEC 62443 and ASHRAE 135.

- Developing repeatable methods for integration and documentation.

You are not just building systems. You are building confidence, continuity, and capability.

## Where the Industry Is Going

- Unified platforms that bring all building systems under one interface.

- Cloud integration for analytics, service, and mobile access.

- AI-based optimization using building data to auto-tune systems.

- More regulation around cybersecurity, carbon reporting, and open access.

- Greater collaboration between trades, IT, and design teams.

Buildings will not just be smart. They will be strategic.

A checklist summarizing future-ready characteristics is included in Appendix K.

## Final Thoughts

This book is not about promoting one protocol, product, or platform. It is about creating a smarter industry. One that treats integration as a baseline, not a bonus. One where practitioners lead with understanding and deliver with precision. One where buildings are easier to operate, safer to use, and better aligned with the organizations they serve.

The tools are here. The standards are available. The opportunity is real.

Now it is your move.

Make the connections. Make the case. Make it work—together.

**Final Summary: A Future-Ready Mindset**

This chapter is not a checklist. It is a mindset.

A future-ready building is not just built for today's technology. It is built to evolve with tomorrow's demands. It is flexible, secure, interoperable, and understandable. But that does not happen automatically. Someone has to lead the conversation. Someone has to connect the dots before they become missed opportunities.

That someone might be you.

Whether you are writing the specification, managing the bid, installing the gear, or supporting it years later, you are not just building systems. You are shaping what smart buildings actually mean in the real world.

Do not wait for the industry to define "smart" on your behalf. Raise the standard.

Keep learning. Stay curious. Speak up early. Ask better questions. Document everything.

Help the next person—because you will not always be the one in the room.

There is no single protocol for this kind of leadership.

But it starts with one move: take integration seriously—
before someone else does not.

# References

McHugh, C. A. (2025, January 14). *You cannot have a smart building in a dumb city: The integration imperative.* LinkedIn. https://www.linkedin.com/pulse/copy-you-cannot-have-smart-building-dumb-city-christine-a--rsp9e/

Zetter, K. (2013, May 1). Researchers hack building control system at Google headquarters. Wired. https://www.wired.com/2013/05/googles-control-system-hacked/

Cupertino Electric Inc. (2015, September 21). Cyber security: Target's 2013 data breach. Cupertino Electric Inc. https://www.cei.com/about-cei/media-room/blog/cyber-security-targets-2013-data-breach

# Appendices

# Appendix A: Names and Acronyms in Building Controls

In the world of building systems, terminology often varies by trade, vendor, region, or even decade. A term that means one thing to an HVAC technician might mean something entirely different to a lighting integrator or IT professional.

This appendix serves as a quick-reference guide for decoding the most common names and acronyms you will encounter—especially useful when trying to reconcile what is being called "the front-end," "the BAS," or "the BMS server."

**General System Names**

- BAS: Building Automation System

- BMS: Building Management System (common in Europe and Asia)

- EMS: Energy Management System (may focus on metering or HVAC performance)

- EMCS: Energy Management and Control System (used in U.S. government specs)

- UMCS: Utility Monitoring and Control System (U.S. military and federal use)

- DDC: Direct Digital Control (refers to controller-based architecture)

- IBMS: Integrated Building Management System (multi-subsystem integration)

- FMS: Facilities Management System (may include maintenance/work orders)

- SCADA: Supervisory Control and Data Acquisition (common in industrial and utility contexts)

- BEMS: Building Energy Management System (energy efficiency focused, Europe)

- iBAS: Intelligent Building Automation System

- EPMS: Electrical Power Monitoring System (used in critical power environments)

- BOS: Building Operating System (newer term for cloud-based supervisory platforms)

## Subsystem and Integration Terms

- HMI: Human-Machine Interface (often used in industrial contexts)

- GUI: Graphical User Interface (the "dashboard" or "front-end")

- OPC / OPC UA: Communication standard for interoperability (especially in industrial controls)

- IoT: Internet of Things (broad term for connected sensors and devices)

- API: Application Programming Interface (used for software-to-software integration)

- FDD: Fault Detection and Diagnostics

- VFD: Variable Frequency Drive (motor control device)

- PID: Proportional-Integral-Derivative (control loop algorithm)

- RTU: Remote Terminal Unit (legacy SCADA field device)

- PLC: Programmable Logic Controller (used in industrial or utility applications)

- MS/TP: Master-Slave/Token-Passing (BACnet over RS-485)

- BTL: BACnet Testing Laboratories (certification for BACnet devices)

**Protocol Acronyms**

- BACnet: Building Automation and Control Network (widely used open protocol)

- Modbus: Industrial protocol developed by Modicon; still common in HVAC and utility systems

- KNX: European fieldbus standard, popular in lighting and shading systems

- DALI: Digital Addressable Lighting Interface

- LON / LonWorks: Legacy protocol used in some BAS installations

- MQTT: Lightweight publish/subscribe protocol for IoT

- REST: Web API communication style (stateless, used in modern integrations)

- SNMP: Simple Network Management Protocol (common in IT for monitoring network devices)

**Final Note**

This list is not exhaustive, but it covers the terms most likely to appear in specifications, proposals, vendor materials, or during interdisciplinary coordination.

# Appendix B: Locations of Controls in U.S. Construction Specifications

In the U.S., building automation and control systems can be referenced in numerous locations within a construction specification, often formatted in accordance with the CSI MasterFormat system. These references span across multiple divisions and must be carefully considered and cross-referenced during design, bidding, and construction. Misalignment or omission can lead to costly change orders or scope confusion.

**Division 01 – General Requirements**

- **01 33 00 – Submittal Procedures**: Governs the submission of control system documentation, shop drawings, sample programs, and integration details.

- **01 78 00 – Closeout Submittals**: Encompasses requirements for O&M manuals, record drawings, system backup files, and training records.

- **01 91 00 – Commissioning Requirements**: Includes full commissioning scope, detailing roles, reporting formats, performance expectations, and integrated testing protocols.

**Division 12 – Furnishings**

- **12 24 00 – Window Shades**: Specifications for automated or motorized shading systems, especially when integrated with lighting or daylighting control strategies. May include fabric type, motor control, and integration expectations with BMS or AV.

**Division 23 – HVAC**

- **23 09 00 – Instrumentation and Control for HVAC**: The core specification for HVAC control system components, covering sensors, control panels, logic diagrams, and functional intent.

- **23 09 23 – Direct Digital Control (DDC) System for HVAC**: Detailed language around DDC platforms, controller hierarchy, software environment, and interface standards.

- **23 09 93 – Sequence of Operations for HVAC Controls**: Explicitly details how each piece of mechanical equipment should operate, react to conditions, and integrate with the broader system.

**Division 25 – Integrated Automation**

- **25 05 11 – Cybersecurity for Facility Controls**: Defines secure system architecture, credentials, authentication policies, remote access methods, patching responsibilities, and compliance with NIST or DoD standards.

- **25 10 00 – Integrated Automation Network Equipment**: Addresses communication infrastructure, BACnet routing, gateways, switches, and device segmentation.

- **25 10 10 – Utility Monitoring and Control System (UMCS) Front End**: Found on U.S. government and military projects, this section outlines the supervisory layer and interface standardization for multiple base-wide systems.

- **25 30 00 – Integrated Automation Instrumentation and Terminal Devices**: Includes field-mounted sensors, devices, and termination practices.

- **25 50 00 – Integrated Automation Facility Controls**: Addresses high-level integration architecture, enterprise dashboards, data sharing expectations, and intersystem synchronization.

Note: Not all projects use Division 25—smaller or design-bid-build projects may specify controls solely under Division 23 or within each discipline.

### Division 26 – Electrical

- **26 09 23 – Lighting Control Devices**: Includes relay panels, dimmers, wall stations, and occupancy sensors tied to electrical rooms.

- **26 09 43 – Network Lighting Controls**: Specifies digital lighting control networks, addressing, zone programming, and integration with daylight sensors or BAS.

### Division 27 – Communications

- **27 15 00 – Communications Cabling**: Encompasses structured cabling that may support BACnet/IP, PoE controllers, AV equipment, and remote control panels.

- **27 21 00 – Data Communication Network Equipment**: Includes IT equipment and network segmentation that supports operational technology (OT) infrastructure.

**Division 28 – Electronic Safety and Security**

- **28 13 00 – Access Control**: Devices such as door controllers, readers, and associated network interfaces.

- **28 23 00 – Video Surveillance**: IP-based video systems, integration with event alarms, and streaming requirements.

- **28 31 00 – Fire Detection and Alarm**: Reference to FACP behavior, fireman's override, monitoring point availability, and relays for smoke control.

**Division 33 – Utilities**

- **33 49 00 – Control and Instrumentation for Utility Systems**: Covers metering, pumping station automation, stormwater monitoring, lift stations, and other infrastructure often connected to SCADA or UMCS platforms.

The presence of control-related content in so many locations requires tight coordination during both design and bidding. Controls engineers must work closely with architects, MEP engineers, and general contractors to ensure all systems are accounted for, properly referenced, and cleanly integrated. Failure to do so often results in missed equipment, communication gaps, and delays in commissioning. An organized and holistic understanding of specification division placement is a key skill for every controls practitioner.

# Appendix C: Trades and Contractor Titles Associated with Building Controls

The responsibility for implementing, installing, programming, and maintaining building control systems may fall to a variety of professionals and trade contractors, depending on project delivery method, region, system complexity, and organizational structure. Below is a list of common titles that may be responsible for control system work.

### Design and Engineering Roles

- **Mechanical Engineer (MEP Engineer)** – Typically responsible for specifying HVAC controls.

- **Electrical Engineer** – May design lighting controls and power for automation panels.

- **Controls Design Engineer** – Specializes in layouts, sequences, point lists, and integration strategies.

- **Technology Designer** – Covers AV, IT, and specialty system controls.

### Installation and Implementation Contractors

- **Temperature Controls Contractor** – Traditional installer of HVAC control systems, often DDC integrators.

- **Building Automation Contractor** – May include both HVAC and lighting control responsibilities.

- **Systems Integrator** – Focuses on unifying multiple systems (e.g., HVAC, lighting, access control).

- **Low-Voltage Contractor** – May install cabling and devices for BAS or security systems.

- **Electrical Subcontractor** – Installs relays, conduit, and some sensors; often supports lighting controls.

- **Mechanical Subcontractor** – May install actuators, sensors, and interface components as part of equipment installs.

- **Fire Alarm Contractor** – Installs devices with FACP integration to BAS.

## Owner and Facility Roles

- **Building Engineer** – Operates and maintains control systems post-occupancy.

- **Energy Manager** – Uses BAS/EMS platforms for energy analytics and optimization.

- **Facility Manager** – Oversees multiple contractors and verifies system performance.

- **Commissioning Agent (CxA)** – Tests and verifies functionality of all systems and integrations.

## Specialized and Support Roles

- **Cybersecurity Consultant** – Ensures control systems follow IT/OT security policies.

- **Network Infrastructure Contractor** – Installs and configures switches, VLANs, and firewalls for BAS.

- **AV Integrator** – Designs and implements integrated audiovisual and shading control systems.

- **Shade & Blind Contractor** – Provides and connects motorized shade systems to the BAS.

## Government/Military Titles (UMCS/EMCS)

- **UMCS Integrator** – Specializes in Utility Monitoring and Control Systems.

- **Energy Monitoring Specialist** – Tracks performance across distributed facilities.

- **Base-wide Controls Contractor** – Responsible for enterprise systems across multiple buildings.

Because control systems span multiple trades, careful coordination and a clear scope of work for each party is essential to avoid overlap, gaps, or missed responsibilities. Roles may vary depending on procurement method, local code enforcement, and system manufacturer preferences.

# Appendix D: Prioritized Documentation – Project Initiation & Requirements Phase

*(for Integrated Fire Alarm, BAS, AV, Access Control, and ESS)*

## 1. Project Charter / Scope Document

Defines the high-level goals, budget, schedule, and stakeholder responsibilities.
Clearly outlines the scope of each system, integration expectations, and the division of responsibilities (who owns what: general contractor, low-voltage contractor, security integrator, IT, etc.).
Identifies the authority for decisions across multiple trades and disciplines.

---

## 2. System Architecture Overview (Integrated View)

A high-level diagram showing all major systems and how they interconnect across Purdue levels.
This should include:

- BAS and fire alarm interaction (for example, shutdowns and fan control)

- AV integration with access control (such as room scheduling and occupancy sensors)

- ESS integration with BAS (for example, HVAC setback when unoccupied)

- IT/OT segmentation, remote access, and cloud services

The architecture overview provides early visibility of data paths, network zones, and security domains.

---

## 3. Requirements Specification (Unified and System-Specific)

Unified requirements should cover redundancy, cybersecurity, remote access, disaster recovery, and uptime Service Level Agreements (SLAs).

System-specific requirements include:

- **Fire Alarm:** NFPA/UL standards, monitoring, and response logic

- **BAS:** Environmental control, energy efficiency, time scheduling

- **AV:** User interfaces, control logic, event automation

- **Access Control:** Door hardware, credential types, anti-passback, emergency override

- **ESS:** Surveillance resolution, recording duration, analytics, alarm triggers

Cybersecurity must be integrated here, addressing user roles, authentication, audit trails, firmware patching, VLAN segmentation, and related measures.

---

## 4. Integration and Interoperability Requirements

Protocols and APIs should be clearly defined, such as BACnet, ONVIF, SIP, Modbus, OPC UA, RESTful APIs, and MQTT.

Define key elements:

- Points of interface (which system pushes or pulls data)

- Shared time synchronization method (such as NTP hierarchy)

- Unified naming and tagging conventions

- Event correlation logic (for example, a fire alarm triggering automatic access control door unlocks)

## 5. Vendor and Technology Standards

Define approved platforms for each system (for example, Johnson Controls for BAS, Lenel for Access Control). Establish firmware and software baselines, network switch standards, UPS requirements, and enclosure types.

Cybersecurity standards must include:

- Password policies

- Patch management procedures

- Antivirus compatibility

- Logging and auditing tools

**Optional (But Smart to Do Early)**

**Stakeholder Roles Matrix (RACI):**
Defines who is Responsible, Accountable, Consulted, and Informed across systems and integration tasks.

**Design Coordination Requirements:**
Defines how trades will coordinate risers, pathways, rack space, grounding, and equipment placement.

# Appendix E: Integration Matrix: Sample System Interactions

The following examples illustrate common interactions between building systems. These examples are valuable during design coordination, construction, and commissioning.

**Sample System Interactions**

- Fire Alarm to HVAC: Shut down air handlers when a fire alarm activates. Typically achieved through dry contact or BACnet. This is required by NFPA 72.

- Fire Alarm to Access Control: Release door locks upon alarm activation. Usually triggered by relay output. Egress must prioritize fail-safe behavior, especially on magnetic locks.

- Access Control to Lighting: Turn on entry lighting after a valid badge read during after-hours periods. Typically integrated through BACnet or an API connection to enhance occupant safety and comfort.

- Access Control to HVAC: Set the HVAC zone to occupied mode when an occupant badge is scanned. This is often done over BACnet/IP and supports energy-saving automation.

- AV System to Lighting: Adjust lights to a preset scene when a presentation mode is selected. Triggered through an API call or contact closure from the AV processor.

- AV System to Shades: Lower blackout shades when the video projection scene is activated. Typically integrated using BACnet, APIs, or simple contact closure to prevent glare.

- Daylight Sensor to Lighting: Dim or turn off lights when daylight levels exceed a predefined threshold. Integration is typically through DALI or BACnet for daylight harvesting strategies.

- Occupancy Sensor to HVAC and Lighting: Set a zone to occupied status when motion is detected. The sensor output must be shared appropriately, often via BACnet, and occupancy logic must be jointly owned.

- Intrusion Detection System to HVAC: Set the zone to unoccupied when an alarm system is armed. This is typically integrated through an API or relay output to trigger energy conservation actions.

- BAS (HVAC) to AV System: Pre-cool event spaces when a scheduled meeting or event is triggered by a calendar API. Integration often happens through BACnet or REST API, improving both comfort and AV system readiness.

- Utility Meter to BAS: Load shedding or alert generation when peak demand is exceeded. Usually integrated using Modbus or pulse input, and may tie into demand response programs.

- Generator Monitoring to BAS: Display generator run status and alarms. Data is collected through Modbus RTU or Modbus IP for operator monitoring. Typically read-only for safety.

- BAS to Enterprise Dashboard: Provide live and trended control points for portfolio-level analytics and reporting. Data is shared through MQTT, API, or BACnet/IP, with consistent point naming crucial for reliable reporting.

**Integration Matrix Usage Tips**

- Treat this matrix as a living document. Update it throughout the design, construction, and commissioning phases.

- Add responsibility assignments or test status columns during construction.

- Include a working version of this matrix as part of the Division 25 specification or project appendix.

- Keep integration language neutral. Focus on the system function, not on vendor-specific marketing terms.

# Appendix F: Sample Division 25 Spec

A **sample Division 25 spec section** specifically focused on **Integration Matrix Coordination**. This can be placed as **Section 25 05 01 – System Integration Coordination**. This is followed by **Section 25 10 10 – Integrated Automation: BAS Front End and Integration**.

---

## SECTION 25 05 01 — System Integration Coordination

### PART 1 – GENERAL

### 1.1 SUMMARY

A. This Section outlines the integration coordination requirements between various building systems, including but not limited to HVAC, lighting, fire alarm, access control, audiovisual, metering, and BAS.
B. This Section provides a standardized format for identifying system interaction points and assigning responsibility for implementation and commissioning.
C. All integrations shall follow the Integration Matrix shown in Table 25 05 01-1.

### 1.2 RELATED SECTIONS

- Section 25 10 10 – BAS Integration

- Section 23 09 00 – Instrumentation and Control for HVAC

- Section 26 09 23 – Lighting Control

- Section 27 15 00 – Communications Cabling

281

- Section 28 13 00 – Access Control

- Section 28 23 00 – Video Surveillance

- Section 28 31 00 – Fire Alarm

- Section 01 91 00 – Commissioning

## 1.3 SUBMITTALS

A. **Integration Matrix Plan**: Submit a draft matrix based on Table 25 05 01-1 with project-specific values and assigned parties.

B. **Test Plans**: Submit test scripts or verification plans for each integration item.

C. **As-Built Integration Matrix**: Provide final matrix as part of project closeout documentation.

## 1.4 COORDINATION

A. Assign an **Integration Coordinator** responsible for tracking, updating, and verifying each matrix item.

B. Each contractor must cooperate with the Integration Coordinator and provide necessary labor and programming to complete integration points.

C. Integration work shall not be considered complete until it has passed commissioning and the final matrix is verified.

---

## PART 2 – PRODUCTS

*Not Used.*

---

## PART 3 – EXECUTION

### 3.1 INTEGRATION MATRIX

Integration interactions must be implemented and tested according to the following descriptions. Add or revise integration points as needed to match the specific project scope.

**Integration Interactions**

- Fire Alarm to HVAC: Shut down air handlers when a fire alarm is active. Triggered by dry contact or BACnet communication. Coordination responsibility: Fire Alarm contractor and HVAC contractor.

- Access Control to Lighting: Turn on lighting upon a valid card swipe. Triggered through API or BACnet connection. Coordination responsibility: Access Control contractor and Lighting contractor.

- Access Control to HVAC: Enable occupied mode when entry is detected at a secured door. Triggered through BACnet/IP communication. Coordination responsibility: Access Control contractor and Building Automation System (BAS) team.

- AV System to Lighting: Adjust lighting scenes automatically when presentation mode is selected. Triggered through API call or contact closure. Coordination responsibility: AV contractor and Lighting contractor.

- AV System to Shades: Lower blackout shades when the projector is turned on. Triggered through API or BACnet integration. Coordination responsibility: AV contractor and Shade Control vendor.

- Daylight Sensor to Lighting: Adjust brightness levels when measured lux exceeds the defined setpoint. Triggered through DALI or BACnet communication. Coordination responsibility: Lighting contractor.

- Intrusion Detection System to HVAC: Set zone to unoccupied when the security alarm is armed. Triggered through dry contact or API integration. Coordination responsibility: Security contractor and BAS team.

- Utility Meter to BAS: Initiate load shedding logic upon a peak demand event. Triggered through Modbus or API integration. Coordination responsibility: BAS team and Electrical contractor.

- Generator Controller to BAS: Display generator operational status when the generator is running. Triggered through Modbus or BACnet communication. Coordination responsibility: Electrical contractor and BAS team.

---

## 3.2 TESTING

A. Perform integration testing using a scripted checklist approved by the Owner's representative.

B. Testing shall verify both physical signals (contacts, relays) and logical sequences (network-based triggers).
C. All test results shall be documented and signed by all responsible trades.

## 3.3 FINAL ACCEPTANCE

A. Final acceptance of system integration is contingent upon verified completion of all matrix items.
B. Owner reserves the right to require re-testing or reprogramming if performance is inconsistent or undocumented.

## SECTION 25 10 10 – Integrated Automation: BAS Front End and Integration

## PART 1 – GENERAL

### 1.1 SUMMARY

A. This section covers the provision, configuration, and implementation of a Building Automation System (BAS) front-end platform and the integration of subsystems using open standard protocols (BACnet, Modbus, OPC, etc.).
B. Includes the supervisory control software, user interface, licensing, and all integration programming.
C. Coordinates with multiple building systems including HVAC, lighting, electrical metering, life safety, access control, video surveillance, and others as indicated.

## 1.2 RELATED SECTIONS

- 23 09 00 – Instrumentation and Control for HVAC

- 26 09 23 – Lighting Control Systems

- 27 15 00 – Communications Cabling

- 28 13 00 – Access Control

- 28 23 00 – Video Surveillance

- 28 31 00 – Fire Alarm

- 25 05 01 – System Integration Coordination

- 25 05 11 – Cybersecurity for Control Systems

---

## 1.3 SUBMITTALS

A. Product Data – Manufacturer's cut sheets, technical specifications, software architecture, supported protocols.
B. Integration Plan – Description of all third-party system integrations, including method (BACnet/IP, Modbus TCP, API), data points, triggers, and sequencing.
C. Network Architecture – Diagram showing IP assignments, VLANs, gateways, firewalls, and host locations.
D. Sequence of Operations – For each integrated system, including mode transitions and triggers.
E. Commissioning Checklist – Based on Appendix C of this manual or custom matrix provided.

---

## 1.4 QUALITY ASSURANCE

A. Installer must be factory-authorized and experienced in control system integration for no fewer than 5 similar projects.

B. All software licenses shall be perpetual, non-expiring, and assigned to the Owner.

C. All protocols shall be fully open-standard (e.g., BACnet B-AWS, Modbus, OPC UA) and not rely on proprietary formats.

---

## 1.5 COORDINATION

A. Coordinate all integration points with other Division 21–28 contractors.

B. Integration responsibility shall be assigned per the Integration Matrix in Section 25 05 01.

C. Participate in commissioning meetings and verification testing for all integration paths.

---

## PART 2 – PRODUCTS

### 2.1 BAS SOFTWARE PLATFORM

A. Provide a web-based supervisory control platform that supports:

- Graphical navigation between systems and floors

- Real-time trending, alarming, and point overrides

- Scheduling with holiday calendar support

- Secure user login with role-based access

- Native support for BACnet/IP, Modbus TCP/IP, and OPC UA

B. Platform must be hosted on an Owner-provided virtual machine or a secure appliance as approved.

C. Software shall be licensed to the Owner with all necessary configuration tools, programming environments, and drivers delivered.

## 2.2 NETWORK INTERFACES AND PROTOCOLS

A. All systems shall communicate over a shared IT-managed network unless otherwise approved.
B. Use of protocol gateways shall be limited and must be documented in the Integration Plan.
C. All BACnet devices shall be BTL Listed and conform to ASHRAE 135 standards.
D. Modbus points shall follow standard addressing tables and use Floating Point or Integer formats unless specified otherwise.

## PART 3 – EXECUTION

### 3.1 INSTALLATION

A. Mount head-end servers, network switches, and controllers per approved network diagrams.

B. Provide all patch cables, labeling, IP addressing, and port configuration in coordination with the Owner's IT team.

## 3.2 PROGRAMMING

A. Implement all sequences of operations as documented.
B. Create user interface graphics for each major system with drill-down navigation.
C. Program alarms, schedules, and user roles per Owner's operational structure.

## 3.3 TESTING AND COMMISSIONING

A. Verify each integration path using documented field commissioning scripts.
B. Demonstrate operation of each system and sequence to the Owner.
C. Submit final as-built documentation including:

- Integration Matrix with actual values

- Export of all trend logs

- Login credentials and licensing keys

- Editable graphics files and point databases

## 3.4 TRAINING

A. Provide 8 hours of Owner training, including:

- Navigation and control of the BAS front end

- Troubleshooting and alarm response

- Scheduling and setpoint changes

- Exporting reports and trends

---

**3.5 WARRANTY**

A. One-year warranty covering all hardware, software, configuration, and labor.
B. Include a support contact with guaranteed response time and remote access capabilities.

---

This is a **Division 25 specification snippet** you can include in either **Section 25 05 11 (Cybersecurity for Control Systems)** or as an **addendum to Section 25 10 10**. It outlines expectations for using a **password vault** during commissioning and ongoing system maintenance, in line with cybersecurity best practices and NIST/ISA 62443 guidance.

**Password Vaulting Requirements – Spec Snippet (Division 25)**

**PART 1 – GENERAL**

**1.1 SUMMARY**

This section outlines the requirements for **credential management and password storage** for all integrated systems included in the building automation and control scope of work. A centralized, secure password vault shall be used to manage all system access credentials provided or created during design, programming, installation, and commissioning.

## 1.2 SUBMITTALS

A. **Password Vault Implementation Plan**, including:

- Selected vault platform (cloud-based or local)

- User access hierarchy (admin, technician, read-only)

- Description of how credentials will be delivered to Owner

- Backup, restore, and recovery procedures

- Rotation policy (default: annual or upon personnel change)

- Emergency access provisions

B. **Vault Access Logs and Summary Report** – Submitted at closeout, showing when each password was set, tested, or updated during commissioning.

---

## PART 2 – PRODUCTS

## 2.1 APPROVED VAULTING SOLUTIONS (Minimum Requirements)

- AES-256 encryption at rest and TLS 1.2 or higher in transit

- Role-based access control (RBAC) with audit logging

- Exportable and encrypted backup format

- Optional: AD/LDAP integration

- Acceptable platforms include:

  - **CyberArk**, **HashiCorp Vault**, **Keeper**, **Bitwarden (Teams/Enterprise)**

  - Offline (air-gapped) sites may use **KeePassXC** or equivalent

---

## PART 3 – EXECUTION

### 3.1 VAULT SETUP AND USE

A. Contractor shall store credentials for all devices and systems requiring authentication, including:

- BAS front end

- Field controllers (DDC/PLC)

- Network switches, routers, and firewalls

- API endpoints and cloud integrations

- Web interfaces or applications requiring login

B. No default manufacturer passwords shall remain active on production systems.

C. Shared accounts shall not be used across systems or vendors unless explicitly authorized by the Owner.

## 3.2 COMMISSIONING REQUIREMENTS

A. Commissioning shall include:

- Validation that all vault entries are accurate and up to date

- Demonstration that passwords are required and tested during functional performance testing

- Confirmation of vault recovery and backup procedures

B. Final vault access and credentials must be turned over securely to the Owner via:

- Encrypted digital handoff (preferred)

- Printed sealed copy (secondary backup)

# Appendix G: Sample Integration Sequence – Conference Room Event Mode

This sample sequence illustrates how multiple building systems integrate to support a scheduled event in a corporate conference room.
The scenario demonstrates typical coordination between audiovisual, HVAC, lighting, and shading systems, triggered by a calendar-based occupancy signal.

**Scenario**

An event is scheduled in the main conference room from 1:00 PM to 3:00 PM.
Integration logic enables automatic environmental adjustments before and during the event.

**Event-Based Integration Logic**

Step 1:
Trigger Source: Room calendar schedule
Target System: BAS (HVAC)
Action: Begin pre-cooling 30 minutes prior to the event (at 12:30 PM).

Step 2:
Trigger Source: Room calendar schedule
Target System: Lighting Control System
Action: Turn on general lighting and adjust to a preset scene.

Step 3:
Trigger Source: Room calendar schedule

Target System: AV System
Action: Power on presentation equipment and display a welcome screen.

Step 4:
Trigger Source: AV Touch Panel Pressed
Target System: Shades
Action: Lower blackout shades to prevent glare.

Step 5:
Trigger Source: AV Scene Activated
Target System: Lighting Control System
Action: Adjust lighting to "Presentation" scene with 50 percent perimeter lighting and 30 percent center lighting.

Step 6:
Trigger Source: Occupancy Sensor Inactive
Target System: HVAC and Lighting Systems
Action: After 15 minutes of vacancy, revert HVAC to unoccupied mode and turn off lights.

Step 7:
Trigger Source: Manual Override by Facility Team
Target System: All Systems
Action: Bypass schedule and place the room in a "Meeting Hold" state, holding current settings.

**Sequence Notes**

- Time-Based Triggers: Schedule data is pulled via iCalendar or API integration from room booking software.

- Scene Control: The AV system manages its own scene presets but triggers BACnet or API calls to the lighting and shade systems.

- Fallback Logic: If calendar data is unavailable, occupancy sensors are used to determine zone mode.

- User Override: AV or facilities personnel can override the sequence using a wall station or web interface.

**Why This Matters**

This example helps stakeholders visualize the functional flow of an integrated system and is often used during early coordination meetings or control narrative development. Including these sequences in design-phase documentation helps ensure:

- Proper assignment of integration responsibilities.

- Realistic expectations of system performance.

- Reduction in programming rework during commissioning.

# Appendix H: Sequence of Operations Template (Per System)

**Appendix H: Sequence of Operations Template (Per System)**

This template provides a structured approach for writing control and integration sequences. Each section focuses on describing how a single system responds to internal logic, external triggers, and integration with other systems.
Use this for all major systems (HVAC, lighting, AV, access control, etc.) and adjust based on project scope.

**System Overview**

- System Name:

  _____

- Controlled Zones:

  _____

- Associated Equipment:

  _____

- Primary Function:

  _____

- Control Platform or Protocol:

  _____

**Operating Modes**

Define all modes this system may enter, whether triggered manually, automatically, or via external input.

Example Modes:

- Occupied: Triggered by schedule or occupancy sensor. Normal operating setpoints and control logic.

- Unoccupied: Triggered by schedule or alarm armed signal. Reduced setpoints or shutdown of non-critical loads.

- Standby: Triggered by vacancy delay timer or schedule gap. Maintains conditions with minimal energy usage.

- Override: Triggered by manual switch or touch panel. Holds system in current state until cleared.

- Alarm Mode: Triggered by fire alarm, intrusion, or system fault. Forces shutdown or engages safety sequence, depending on system.

**Integration Triggers (Inbound)**

List any signals this system receives from other systems.

Example Integration Triggers:

- Source: Fire Alarm
  Signal: Alarm active
  Action Taken: Shut down HVAC unit and unlock access-controlled doors.

- Source: Access Control
  Signal: Badge swipe
  Action Taken: Set zone to occupied mode.

- Source: AV Touch Panel
  Signal: "Presentation Scene" selected
  Action Taken: Dim lights and lower shades.

**Integration Responses (Outbound)**

List any signals this system sends to other systems.

Example Integration Responses:

- Target: Lighting System
  Signal: Occupied command
  Triggered When: HVAC enters Occupied mode.

- Target: Energy Dashboard
  Signal: Demand reading
  Triggered When: Every 15 minutes.

- Target: AV System
  Signal: System ready confirmation
  Triggered When: HVAC zone reaches setpoint.

**Control Sequence Narrative**

Provide a clear, step-by-step narrative for how the system operates under normal and exceptional conditions.

Example Narrative:
Upon entry during after-hours, the access control system triggers the HVAC zone to switch to Occupied mode and illuminates corridor lighting to 50 percent.
The BAS verifies that the zone temperature is within setpoint range before sending a "Room Ready" signal to the AV system.

If no occupancy is detected after 15 minutes, the system automatically reverts to Standby mode to conserve energy.

**Failure and Alarm Response**

Describe how the system responds to faults, loss of communications, or emergency signals.

Example Responses:

- Loss of Communication: System reverts to safe default state and generates an alert.

- Fire Alarm Trigger: Immediately shut down mechanical equipment and unlock access-controlled doors.

- System Fault Detected: Isolate affected subsystem and generate operator alert.

- Manual Emergency Stop Activated: Override all automatic functions and force shutdown.

**Commissioning Notes and Test Instructions**

- Verify all mode transitions such as Occupied, Unoccupied, Standby, and Override under field conditions.

- Test each integration trigger individually and validate that the correct response occurs.

- Confirm that time delays, interlocks, and manual overrides function as designed.

- Document all observed behaviors and note any deviations during functional performance testing.

# Appendix I: Division 28 + Division 25 Interface Summary

## Purpose of This Summary

As modern buildings increasingly demand integrated control of security, HVAC, lighting, and other subsystems, it is critical to understand how Division 28 (Electronic Safety and Security) interacts with Division 25 (Integrated Automation).

This appendix provides an overview of key integration points, system boundaries, and applicable standards to guide coordination between trades and ensure a secure, maintainable, and code-compliant installation.

## Key Integration Touchpoints

### Access Control (28 13 00)

Division 25 Role: BAS triggers HVAC or lighting based on badge events.

Typical Integration Method: BACnet/IP, API, or relay contact.

Purpose: Energy savings and occupancy-based control.

### Intrusion Detection (28 16 00)

Division 25 Role: BAS shifts system to setback or unoccupied mode when armed.

Typical Integration Method: Digital input or API.

Purpose: Nighttime energy saving and threat response.

### Video Surveillance (28 23 00)

Division 25 Role: Event-based integration such as motion or

alarm triggers.

Typical Integration Method: API or contact output to BAS or lighting systems.

Purpose: Coordinated lighting or alerting based on video events.

### Fire Alarm (28 31 00)

Division 25 Role: Emergency sequences like AHU shutdown, door unlock, and egress lighting control.

Typical Integration Method: Dry contact or BACnet/IP.

Purpose: Life safety integration.

### Mass Notification and Paging Systems

Division 25 Role: BAS triggers pre-recorded emergency messages based on sensor data.

Typical Integration Method: API or contact closure.

Purpose: Emergency communication tied to occupancy or alarm states.

### Applicable Standards by Function

### Secure Access Reader Communication

Standard: SIA OSDP (Open Supervised Device Protocol).

Purpose: Replaces legacy Wiegand protocols with encrypted, bi-directional access control signaling.

### Access Control Equipment

Standard: UL 294.

Purpose: Listing required for commercial compliance and insurance acceptance.

### Alarm and Sensor Devices

Standard: UL 1076 and NEMA SB 30.

Purpose: Establishes performance expectations and false alarm prevention measures.

**Video Surveillance Protocols**
Standard: IEC 62676.
Purpose: Defines video streaming, recording, and interoperability formats.

**Cybersecurity for Networked Devices**
Standards: NIST SP 800-82 and ISA 62443.
Purpose: Protects operational technology (OT) systems from cybersecurity threats.

**Organizational Governance and Risk Management**
Standard: ASIS Physical Security Risk Assessment.
Purpose: Aligns enterprise-level security goals with technical implementation strategies.

**Design and Coordination Best Practices**

- Assign system boundaries early. Define which contractor is responsible for programming, logic, and network connections.

- Use an Integration Matrix. Include all Division 28 system interactions and clearly define responsible parties.

- Verify protocol compatibility. Confirm that access control and video systems can communicate natively with the BAS.

- Protect credentials. Store all passwords, keys, and access logs in a secure password vault approved for critical infrastructure.

- Commission as a unified system. Integration testing must verify that badge events, intrusion alarms, and fire events trigger proper coordinated responses.

**Example Use Case: After-Hours Entry Sequence**

When a user swipes a badge at the south entrance, the BAS system sets the corresponding HVAC zone to occupied, turns on corridor lighting, and logs the event for energy tracking.
If the intrusion detection system is still armed, it automatically disarms upon successful badge authentication.
All event sequences, user credentials, and scripts are securely stored in a controlled-access password vault to maintain cybersecurity integrity.

**Visual Integration Overview**

(Note: Diagram Placeholder)

The system relationships typically include:

- Access Control to BAS (setting occupancy modes based on badge access)

- Fire Alarm to BAS (forcing emergency modes and shutdowns)

- Intrusion Detection to BAS (triggering nighttime setbacks and security responses)

- BAS to Lighting, HVAC, and AV systems (coordinated environmental response)

- Secure credential management across all integrated platforms

# Appendix J: Niagara vs. Open Frameworks in Division 25 Planning

Choosing the right integration platform has major impacts on long-term flexibility, cost, and maintainability. Below is a text-based comparison of Niagara Framework and Open-Source options, suitable for Division 25 planning.

**Licensing Model**

- Niagara Framework: Proprietary; requires a Tridium license.

- Open-Source Frameworks (e.g., Sedona, Node-RED): Free and open-source under BSD, MIT, or Apache licenses.

**Developer Access**

- Niagara: Limited to certified developers.

- Open Frameworks: Open to all developers, community-supported.

**Integration Capabilities**

- Niagara: Wide support through licensed drivers, including major building systems and protocols.

- Open Frameworks: Broad but depends on community contributions or custom-built integrations.

**Tooling Environment**

- Niagara: Uses Niagara Workbench for development and configuration.

- Open Frameworks: Options include web-based tools (Node-RED), Sedona Workbench, or command-line interfaces.

**Edge Device Support**

- Niagara: Excellent support via JACE and PX controllers.

- Open Frameworks: Good support on devices like Raspberry Pi, BeagleBone, and embedded Linux systems.

**Protocol Support**

- Niagara: Native support for BACnet, Modbus, KNX, SNMP, oBIX, and more.

- Open Frameworks: Varies, but commonly includes MQTT, Modbus, BACnet, REST APIs, and WebSockets.

**User Interface Customization**

- Niagara: Integrated, professional-grade graphical user interfaces.

- Open Frameworks: DIY dashboards or third-party integrations, often requiring more setup effort.

**Security and Encryption**

- Niagara: Built-in TLS encryption, role-based access control, and audit trails.

- Open Frameworks: Security depends heavily on how the platform is configured and hosted.

## Scalability and Enterprise Fit

- Niagara: High scalability, ideal for large sites, campuses, and enterprise portfolios.

- Open Frameworks: Best suited for small to mid-sized projects or experimental deployments.

## Community and Ecosystem

- Niagara: Strong partner ecosystem among certified integrators and vendors.

- Open Frameworks: Broad developer base across IT, OT, and hobbyist communities.

## Ideal Use Case

- Niagara: Best for commercial, multi-system, multi-vendor building automation systems requiring strict reliability, structured support, and regulatory compliance.

- Open Frameworks: Ideal for low-cost, flexible, or R&D environments where custom development is possible and vendor independence is a priority.

**Conclusion**

Niagara excels in managed, high-complexity environments where compliance, reliability, and structured vendor ecosystems are critical.

Open-source frameworks offer powerful flexibility and lower costs but require greater development effort, technical skill, and operational oversight.

Choose based on the project's scale, goals, and the owner's tolerance for custom support versus commercial structure.

# Appendix K: How to Recognize a Future-Ready Building

This quick-reference checklist helps identify whether a project is truly future-ready—or if there are opportunities to improve.

**1. Interoperability**
Systems communicate through open protocols like BACnet, Modbus, or MQTT, minimizing vendor lock-in.

**2. Network-Ready Infrastructure**
VLANs, cabling, and switch ports are reserved and documented for all building systems during design.

**3. Lifecycle Documentation**
As-built drawings, sequences of operation, network diagrams, and credential lists are complete, accurate, and easily accessible.

**4. Cybersecurity Built In**
Follows established frameworks such as NIST SP 800-82 or IEC 62443. Passwords are securely managed, and remote access is controlled.

**5. Modular and Scalable Design**
The infrastructure supports growth—more devices, greater bandwidth, and future upgrades—without needing complete system overhauls.

**6. Change Management Process**
All modifications are logged, tracked, and communicated. Changes are not hidden in field fixes or left undocumented.

### 7. Digital Twin or Real-Time Models

The building systems can be monitored, tuned, and simulated using a front-end platform or digital twin technology.

### 8. Unified Front-End or Dashboards

Operators can view and manage multiple systems—HVAC, lighting, access, AV—from a single, cohesive interface.

### 9. IT and OT Are on the Same Page

Information Technology (IT) and Operational Technology (OT) teams collaborated during design—not just at project turnover.

### 10. Documented Service Strategy

There is a clear plan for backups, firmware patching, support contacts, and regular system health checks.

---

### Quick Tip

If you can check seven or more of these boxes, you are likely looking at a future-ready building.
If not, the gaps point to opportunities for improvement on your next project.

Jacob Jackson is a Certified Automation Professional (CAP) with more than two decades of experience designing, integrating, and leading operational technology (OT) and building automation projects across commercial, government, and institutional sectors. He specializes in connecting HVAC, lighting, security, audiovisual, and energy systems into unified platforms that align with business goals, cybersecurity standards, and long-term operational strategies.

Throughout his career, Jacob has worked with organizations ranging from Fortune 500 corporations to federal agencies and higher education campuses. He is known for his ability to bridge the gap between engineering, IT, and facility operations, helping project teams move from siloed systems to fully integrated smart building solutions.

In addition to his technical expertise, Jacob has contributed to industry standards committees and led workforce development initiatives promoting integration and automation careers. He is also the founder of Logicmaker Intelligence LLC, a consulting and publishing firm focused on operational technology strategy, smart infrastructure, and controls troubleshooting frameworks.

His writing reflects a commitment to practical, real-world solutions that are grounded in engineering fundamentals and lifecycle planning. *Smart Buildings by Design* is his first book, created to help professionals across the built environment understand the value of integration, the risks of vendor lock-in, and the importance of aligning technology infrastructure with long-term building use.

www.ingramcontent.com/pod-product-compliance
Lightning Source LLC
Chambersburg PA
CBHW062155120626
46550CB00012B/1511